Lecture Notes in Networks and Systems

Volume 378

AF868733

Series Editor

Janusz Kacprzyk, Systems Research Institute, Polish Academy of Sciences, Warsaw, Poland

Advisory Editors

Fernando Gomide, Department of Computer Engineering and Automation—DCA, School of Electrical and Computer Engineering—FEEC, University of Campinas—UNICAMP, São Paulo, Brazil

Okyay Kaynak, Department of Electrical and Electronic Engineering, Bogazici University, Istanbul, Turkey

Derong Liu, Department of Electrical and Computer Engineering, University of Illinois at Chicago, Chicago, USA
Institute of Automation, Chinese Academy of Sciences, Beijing, China

Witold Pedrycz, Department of Electrical and Computer Engineering, University of Alberta, Alberta, Canada
Systems Research Institute, Polish Academy of Sciences, Warsaw, Poland

Marios M. Polycarpou, Department of Electrical and Computer Engineering, KIOS Research Center for Intelligent Systems and Networks, University of Cyprus, Nicosia, Cyprus

Imre J. Rudas, Óbuda University, Budapest, Hungary

Jun Wang, Department of Computer Science, City University of Hong Kong, Kowloon, Hong Kong

The series "Lecture Notes in Networks and Systems" publishes the latest developments in Networks and Systems—quickly, informally and with high quality. Original research reported in proceedings and post-proceedings represents the core of LNNS.

Volumes published in LNNS embrace all aspects and subfields of, as well as new challenges in, Networks and Systems.

The series contains proceedings and edited volumes in systems and networks, spanning the areas of Cyber-Physical Systems, Autonomous Systems, Sensor Networks, Control Systems, Energy Systems, Automotive Systems, Biological Systems, Vehicular Networking and Connected Vehicles, Aerospace Systems, Automation, Manufacturing, Smart Grids, Nonlinear Systems, Power Systems, Robotics, Social Systems, Economic Systems and other. Of particular value to both the contributors and the readership are the short publication timeframe and the world-wide distribution and exposure which enable both a wide and rapid dissemination of research output.

The series covers the theory, applications, and perspectives on the state of the art and future developments relevant to systems and networks, decision making, control, complex processes and related areas, as embedded in the fields of interdisciplinary and applied sciences, engineering, computer science, physics, economics, social, and life sciences, as well as the paradigms and methodologies behind them.

Indexed by SCOPUS, INSPEC, WTI Frankfurt eG, zbMATH, SCImago.

All books published in the series are submitted for consideration in Web of Science.

For proposals from Asia please contact Aninda Bose (aninda.bose@springer.com).

More information about this series at https://link.springer.com/bookseries/15179

Hani Ragab Hassen · Hadj Batatia
Editors

Proceedings of the International Conference on Applied CyberSecurity (ACS) 2021

Springer

Editors
Hani Ragab Hassen
Heriot-Watt University
Dubai, UAE

Hadj Batatia
Heriot-Watt University
Dubai, UAE

ISSN 2367-3370 ISSN 2367-3389 (electronic)
Lecture Notes in Networks and Systems
ISBN 978-3-030-95917-3 ISBN 978-3-030-95918-0 (eBook)
https://doi.org/10.1007/978-3-030-95918-0

© The Editor(s) (if applicable) and The Author(s), under exclusive license
to Springer Nature Switzerland AG 2022
This work is subject to copyright. All rights are solely and exclusively licensed by the Publisher, whether the whole or part of the material is concerned, specifically the rights of translation, reprinting, reuse of illustrations, recitation, broadcasting, reproduction on microfilms or in any other physical way, and transmission or information storage and retrieval, electronic adaptation, computer software, or by similar or dissimilar methodology now known or hereafter developed.
The use of general descriptive names, registered names, trademarks, service marks, etc. in this publication does not imply, even in the absence of a specific statement, that such names are exempt from the relevant protective laws and regulations and therefore free for general use.
The publisher, the authors and the editors are safe to assume that the advice and information in this book are believed to be true and accurate at the date of publication. Neither the publisher nor the authors or the editors give a warranty, expressed or implied, with respect to the material contained herein or for any errors or omissions that may have been made. The publisher remains neutral with regard to jurisdictional claims in published maps and institutional affiliations.

This Springer imprint is published by the registered company Springer Nature Switzerland AG
The registered company address is: Gewerbestrasse 11, 6330 Cham, Switzerland

Preface

Cybersecurity has gained in awareness and media coverage in the last decade. Not a single week passes without a major security incident that affects a company, sector, or governmental agencies. The International Conference on Applied Cyber Security 2021 (ACS21) aims to provide a venue where security experts, both in research and industry, can exchange their knowledge and findings on how to secure information systems. We promote cross-disciplinary and multidisciplinary approaches to cybersecurity. Some of the papers presented in this volume constitute good examples of cross-disciplinary approaches.

These proceedings contain 13 original contributions. There is a high diversity in the problems and methodologies, which shows how rich this area is, yet still open to original contributions. It is interesting though to observe the use of machine learning and artificial intelligence throughout this book, consistent with the current trend of increasingly using these techniques in cybersecurity, both to defend and attack assets. More than half of the contributions are about applications of machine learning to accomplish several cybersecurity tasks, such as malware detection, phishing email detection, Botnet detection, and scanning source code for vulnerabilities. For the convenience of readers, we divided this volume into two parts; the first is focused on machine learning applications to cybersecurity, whereas the second groups the other approaches.

Overall, these proceedings review a panoply of recent techniques that can be leveraged to build secure systems. The goal is to share the community knowledge as well as new ideas and concepts that could inspire future generations of cybersecurity researchers and practitioners.

Target Audience

This book is suitable for cybersecurity researchers, cybersecurity practitioners as well as fresh graduates. It is also suitable for artificial intelligence researchers interested in exploring applications in cybersecurity. Prior exposure to basic machine learning concepts is preferable, but not necessary.

Acknowledgements and Thanks

The editors want to express their gratitude to the authors, program committee members, and the reviewers for their great efforts writing, reviewing, and providing insightful feedback. We would also like to thank Varsha Prabakaran and Thomas Ditzinger from Springer for their help and support in the production of this manuscript.

We would like to extend our special thanks to the keynote speakers, Mr. Biju Hameed, Head of Technology Infrastructure Operations, Dubai Airports, and Anthony Lewis Brooks, Associate Professor, Department of Architecture, Design and Media Technology, Aalborg University, Denmark.

Hani Ragab Hassen
Hadj Batatia

Conference Organization

Honorary Chair

Steve Gill — Associate Professor, Head of School of Mathematical and Computer Sciences, Heriot-Watt University, Dubai, United Arab Emirates

General Chair

Hani Ragab Hassen — Associate Professor, Director of the Institute of Applied Information Security, School of Mathematical and Computer Sciences, Heriot-Watt University, Dubai, United Arab Emirates

General Co-chair

Hadj Batatia — Associate Professor, School of Mathematical and Computer Sciences, Heriot-Watt University, Dubai, United Arab Emirates

Steering Committee

Hani Ragab Hassen — Associate Professor, Heriot-Watt University, Dubai, United Arab Emirates

Abdelmadjid Bouabdallah — University of Technology of Compiegne, France

Hadj Batatia — Associate Professor, Heriot-Watt University, Dubai, United Arab Emirates

International Co-chair

Ahcene Bounceur	Associate Professor, Bretagne Occidentale University, France

Organizing Committee

Madjid Merabti	University of Sharjah, UAE
Adrian Turcanu	Assistant Professor, Heriot-Watt University, UAE
Hind Zantout	Associate Professor, Heriot-Watt University, UAE
Mohammad Hamdan	Associate Professor, Heriot-Watt University, UAE
Abrar Ullah	Associate Professor, Heriot-Watt University, UAE
Smitha Kumar	Assistant Professor, Heriot-Watt University, UAE
Ali Muzaafar	Teaching Assistant, Heriot-Watt University, UAE

Contents

Securing and Hardening Information Systems

Vulnerability and Infection Detection using Machine Learning

Android Malware Detection Using API Calls: A Comparison of Feature Selection and Machine Learning Models

Ali Muzaffar[1](✉), Hani Ragab Hassen[1], Michael A. Lones[2], and Hind Zantout[1]

[1] Heriot-Watt University, Dubai, UAE
{am29,h.ragabhassen,h.zantout}@hw.ac.uk
[2] Heriot-Watt University, Edinburgh EH14 4AS, UK
m.lones@hw.ac.uk

Abstract. Android has become a major target for malware attacks due its popularity and ease of distribution of applications. According to a recent study, around 11,000 new malware appear online on daily basis. Machine learning approaches have shown to perform well in detecting malware. In particular, API calls has been found to be one of the best performing features in malware detection. However, due to the functionalities provided by the Android SDK, applications can use many API calls, creating a computational overhead while training machine learning models. In this study, we look at the benefits of using feature selection to reduce this overhead. We consider three different feature selection algorithms, mutual information, variance threshold and Pearson correlation coefficient, when used with five different machine learning models: support vector machines, decision trees, random forests, Naïve Bayes and AdaBoost. We collected a dataset of 40,000 Android applications that used 134,207 different API calls. Our results show that the number of API calls can be reduced by approximately 95%, whilst still being more accurate than when the full API feature set is used. Random forests achieve the best discrimination between malware and benign applications, with an accuracy of 96.1%.

1 Introduction

Smartphones have become an essential part of our daily lives. Today more than 48% of the world's population use smartphones, which adds up to approximately 3.8 billion people [1]. There has been an exponential growth in the number of smartphone users since 2016 when only 33.58% of the world population used smartphones [1]. Smartphone operating systems allow the users to run applications that can be downloaded from various application repositories available online. Among these operating systems, Android holds the major share of 72.18% in the smartphone market, followed by iOS at 26.96% [2].

The typical use of smartphones includes the storage of sensitive information such as text messages, emails, business data and personal files such as images

© The Author(s), under exclusive license to Springer Nature Switzerland AG 2022
H. Ragab Hassen and H. Batatia (Eds.): ACS 2021, LNNS 378, pp. 3–12, 2022.
https://doi.org/10.1007/978-3-030-95918-0_1

and videos. Moreover, most smartphone users connect to the internet and use a variety of different services, including web and location services. Its popularity combined with the sensitive nature of the data stored in smartphones has led to an increase in the spread of Android malware.

Android allows users to download and install applications from its official application store, Google Play Store [3], and third party online stores. The existence of third party application repositories makes the spread of Android malware easier and quicker. Even Google Play Store cannot guarantee the applications listed in their store are free of malware [4].

Traditional anti-malware techniques use signature-based detection. The signature of the file can be anything from a pattern of bytes to the hash of the file. The signature is compared to a known database to detect malware [5]. However, a small modification to the signature, introduced by adding keywords or lines of code, can cause the malware to evade detection. This can prevent signature-based anti-malware from detecting existing and zero-day malware.

To overcome the limitations of signature-based detection, researchers have explored machine learning (ML) based malware detection. This process requires dataset collection, feature extraction using static and/or dynamic analysis, feature engineering and finally training ML models. Static analysis is carried out without running the application. Features are extracted by unpacking the application and mining features from the manifest and source code. One of the earlier static analysis works was by Peiravian and Zhu [6], who trained their ML models on three different feature sets including permissions (130 features), API calls (1,326 features) and a combination of both (1,456 features). They reported an accuracy of 96.88% using a support vector machine (SVM) trained on both API calls and permissions. Arp et al. [7] extracted around 545,000 features using static analysis, and reported an accuracy of 93.9%, also using an SVM. Ma et al. [8] focused on API calls, and extracted features based on API usage, frequency and sequence. Decision tree (DT), deep neural network (DNN) and long short-term memory (LSTM) models all resulted in F1 scores greater than 96%. Jung et al. [9] selected the top 50 API calls used in benign applications and malware to train a random forest model, and reported accuracies ranging from 97% to 99%.

On the other hand, features can be extracted using dynamic analysis. In this case, the application is run on an emulator and features are extracted during the runtime of the application. Afonso et al. [10] used dynamic analysis to extract system call traces and API calls used by the application. They trained a random forest model and reported an accuracy of 96.82%. Xiao et al. [11] also extracted system call traces and used them as natural language to train an LSTM language model, reporting accuracy rates of up to 96.3%. Features from both static and dynamic analysis can also be combined to build classifiers. This is called "Hybrid Analysis". For instance, Saracino et al. [12] used permissions and market information as static features and system calls, and user activity and API calls as dynamic features to train a k-nearest neighbour model. The authors reported a detection rate of 96.9%.

Dynamic analysis can be difficult to carry out because of the computational resources required to run the analysis on an Android virtual device. Therefore most researchers opt for static analysis to extract features for their ML models. Features based on API calls have produced very promising results. Therefore, we used Android API calls that were extracted using static analysis to train ML models. However, there are over 130,000 API calls that can be used by Android applications. This makes it difficult to train models on such a large feature set. To address this, we use several feature selection algorithms, namely mutual information, Pearson correlation coefficient (PCC) and variance threshold to select relevant features in Android malware detection. We then use the reduced sets of features to train SVM, random forest, Naïve bayes, DT and AdaBoost ML models. We conclude with a comparison of the results of models that were trained using different numbers of features that were selected using feature selection algorithms.

Through our comparative study we made the following contributions:

1. We compared how different ML models performed using the complete API call feature set and subsets produced by the three most commonly used feature selection algorithms in the literature.
2. We showed that random forest models perform the best with the full API feature set, reporting an accuracy rate of 95.9%. We also demonstrated that higher accuracy rates can be achieved by using only 5% of the Android API calls rather than the full API calls feature set.
3. In order to reliably evaluate these models, we collected a new, up to date, dataset of Android applications collected from various sources including popular online application stores.

The remainder of the paper is organized as follows. We introduce our framework, dataset, ML models and feature selection algorithms used in Sect. 2, report and discuss our findings in Sect. 3, followed by the conclusion in Sect. 4.

2 Framework Overview

The Android SDK provides programmers with API calls they can use to implement various functionalities in their applications. These functionalities include providing a GUI to the application, using hardware components of the devices and accessing user location, among many others. We crawled the Android API reference page [13] to gather all the API packages available. A Python script was then developed to extract all the API calls used by the applications by matching the API calls package name.

The aim of this study is to reduce the number of API calls used to train ML models while maintaining the detection rates produced by the full API calls feature set. The following is the framework design for the study:

- **Dataset Collection:** We collected a total of 40,000 Android applications from various sources to extract features and train classification models. These were balanced between 20,000 malware and 20,000 benign applications.

- **Feature Extraction (Static Analysis):** We wrote a Python script to extract the API calls used by the applications. *APKTool* [14] was used to reverse engineer the DEX code file to produce smali files. The smali files were then analysed to extract API calls.
- **Feature Selection:** we used Mutual information, PCC and variance threshold to select the most relevant features.
- **Train Models:** we trained ML models on the full API calls feature set and subsets produced by the feature selection methods.

2.1 Dataset

Android applications are released at a rapid pace. Therefore, a relevant and recent dataset is essential for any framework to have any practical use. Data is one of the most important factors in determining the quality of ML models. Unlike many previous studies reported in the literature, we used a balanced, real life and up to date dataset. The applications we used for our dataset were released from 2019 to 2021.

We used VirusShare's [15] most recent Android malware dataset for our malicious applications dataset. We used a total of 20,000 applications from VirusShare. For our benign dataset, we crawled several Android repositories including UpToDown [16], APKMirror [17] and F-Droid [18]. In total, we downloaded 20,000 benign applications. Each application was labelled using VirusTotal [19] reports. In order to prevent false negatives from leaking into our benign datasets, only applications with zero positives were used for the benign dataset.

We ran a static analysis on our dataset to collect the API calls used by each application and built a Boolean dataset. Let $A = \{API_1, API_2, API_3, \ldots API_n\}$ be the complete API set consisting of a total of n number of API calls. Each application's attribute in the Boolean dataset is set A plus the label. We used 0 to indicate the application does not use the API and 1 indicates that the API is used and label is set to 0 for benign and 1 for malware. For example, if $A = \{API1, API2, API3\}$ and an application used $API1$ and $API3$ and is a malware, the vector of this application will be $M = \{1, 0, 1, 1\}$.

2.2 ML Models

We trained models using Support Vector Machine (SVM), Decision Tree (DT), Random Forest (RF), Naïve Bayes (NB) and AdaBoost on all the feature sets including the complete API calls feature set and the subsets produced by feature selection methods. These are all standard ML models, but for reference we include a brief description of each:

- **Support Vector Machines (SVM)** find the optimal hyperplane that separates the samples from two classes in their n-dimensional feature space. SVMs can also solve non-linear problems by using a kernel trick to project the data into a higher-dimensional space, but here we use an SVM with a linear kernel. SVMs are known for their speed and robustness.

- **Decision Trees (DT)** are tree-structured decision-making processes. Each node in the tree considers a single feature, and based on its value, passes control to one of its two sub-branches. When it reaches the leaf nodes of the tree, a sample is assigned to a particular class. DTs are relatively interpretable ML models, but the use of perpendicular decision planes can limit their accuracy.
- **Random Forest** is another tree-based classifier which uses many DTs to improve the accuracy of single DT models. Random forests train multiple DTs and then output the majority classification. The number of trees we used in our models was 500.
- **Naïve Bayes** is a simple probabilistic classifier model based on Bayes' theorem. It can be used for both classification and regression. Training and testing a Naïve Bayes classifier is comparatively fast in comparison to other ML models, which allows it to scale to large data sets.
- **AdaBoost,** or adaptive boosting, is a classic ensemble learning approach that combines multiple weak classifiers into one relatively strong classifier. For our models we used DT as the base classifier, with maximum number of estimators set to 50.

We evaluate the performance of the models using 10-fold cross-validation (CV) on our dataset. We report the mean and standard deviation of the accuracy, precision, F1-score, true positive rate (TPR) and true negative rate (TNR) to provide a complete picture of how the model performs.

2.3 Feature Selection Methods

Feature ranking is used in machine learning to measure the relevance of the feature to its class label. This helps in selecting the most relevant and informative features in order to improve the model's performance [20]. We used three feature selection algorithms to reduce the number of features from the API calls feature vector: mutual information, variance threshold and PCC.

- **Mutual Information** is the measure of information obtained between two random variables. The value of mutual information is always greater than or equal to zero. Two variables are independent when the value is zero, and, the greater the value, the stronger their relationship is. We calculated the mutual information of all the features in our dataset. Features with the highest mutual information were then used to train the ML models.
- **Variance Threshold** measures the variance of each feature within a dataset, and then eliminates those which have a variance below a specified threshold. It is based on the premise that features which have similar values within different samples tend not to be useful for classification. For instance, in the extreme case where a feature has zero variance, the feature's value is the same for every sample, and therefore provides no information. We calculated the variance of each feature in our dataset. We then used the features with highest variance to train the ML models.

- **Pearson Correlation Co-efficient (PCC)** is used to calculate how linearly dependent a variable is to its target label. The resulting coefficient is greater or less than 0 if the two variables are related, whereas the coefficient is 0 if there is no correlation. We calculated the PCC of all the features in our dataset. The features with the highest PCC were then used to train the ML models.

3 Experimental Results and Analysis

In this section we discuss the ML models and feature selection methods used and report the results from each experiment.

Table 1. Average and standard deviation of evaluation metrics of full API feature set trained on five ML models using 10-fold CV

Classifier	Accuracy	Precision	F1-Score	TPR	TNR
SVM	0.955 ± 0.002	0.957 ± 0.004	0.955 ± 0.002	0.953 ± 0.004	0.957 ± 0.004
DT	0.940 ± 0.000	0.938 ± 0.003	0.941 ± 0.002	0.943 ± 0.002	0.938 ± 0.002
Random Forest	**0.959** ± 0.001	**0.960** ± 0.001	**0.959** ± 0.002	**0.957** ± 0.001	**0.960** ± 0.002
Naïve Bayes	0.744 ± 0.002	0.673 ± 0.003	0.789 ± 0.002	0.957 ± 0.002	0.531 ± 0.005
AdaBoost	0.943 ± 0.002	0.943 ± 0.001	0.943 ± 0.002	0.944 ± 0.004	0.942 ± 0.001

Table 1 shows the results of applying the five ML models to the complete feature set, which comprises 134,207 API calls. It can be seen that random forest achieved the best results in all the metrics, followed by SVM. AdaBoost performed slightly better than DT. Although Naïve Bayes took the least time to train, the accuracy of the model was very low.

We then applied the feature selection algorithms, in order to reduce the number of API calls used for classification. Specifically, the three feature selection algorithms were used to reduce the feature set to sizes between 100 and 30,000 API calls. The results are shown in Fig. 1. This shows that the feature set size does have a significant affect on the accuracy of the ML models. However, beyond a certain feature set size threshold, for most of the models the accuracy approaches those achieved using the full feature set.

Random forest remains the most accurate classifier, regardless of the feature set size. Table 2 shows the detailed metrics for the best performing random forest model, for each of the feature selection methods. The overall best models are found when variance threshold and mutual information are used, with PCC resulting in significantly poorer models. Notably, the model produced after variance thresholding is better in all metrics than the model produced from the full feature set, despite using only 5% of the features.

In general, there was a considerable reduction in time taken to train the ML models on the reduced feature sets. This however did not affect the model accuracy considerably. In fact, SVM for the most part maintained the accuracy it

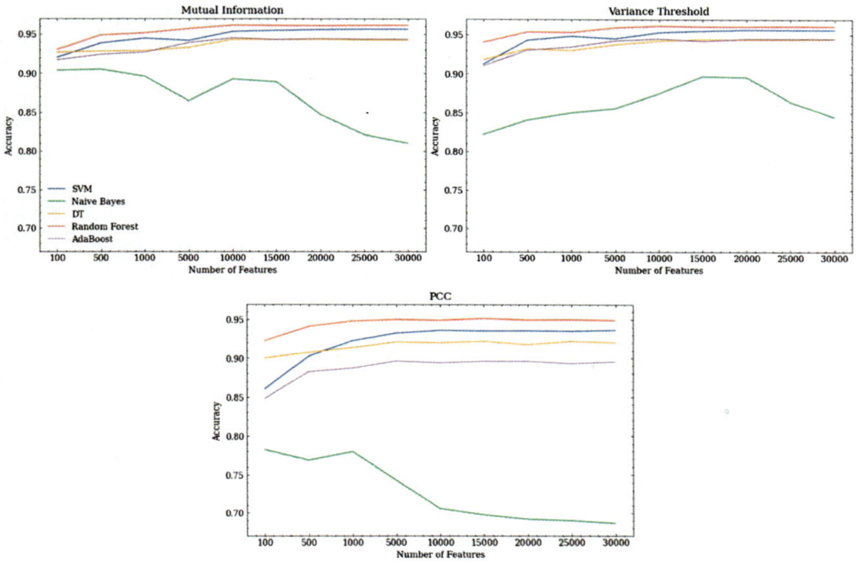

Fig. 1. Mean accuracy of models trained on features selected using mutual information, variance threshold and PCC using 10-fold CV

Table 2. Average and standard deviation of evaluation metrics of the best performing ML models using top features from feature selection algorithms

Feature selection	Classifier	No. of features	Accuracy	Precision	F1-score	TPR	TNR
Mutual information	Random Forest	10,000	0.962 ± 0.001	0.960 ± 0.001	0.961 ± 0.002	0.959 ± 0.002	0.963 ± 0.002
Variance threshold	Random Forest	6,443	0.961 ± 0.002	0.964 ± 0.001	0.961 ± 0.002	0.958 ± 0.002	0.964 ± 0.001
PCC	Random Forest	15,000	0.951 ± 0.001	0.952 ± 0.002	0.951 ± 0.002	0.938 ± 0.001	0.965 ± 0.002

achieved with the full API feature set and even reported higher accuracy with mutual information and variance threshold. AdaBoost only performed better than DT in some feature set sizes when using mutual information and variance threshold, and performed worse than DT in PCC. Naïve Bayes showed an improvement in accuracy with less number of features only when mutual information and variance threshold were used. The accuracy rate increased from 74.4% using full feature set to 90.5% when using mutual information. However, the accuracy dropped considerably as the number of features was increased.

We also observed that there is a high degree of overlap between the features selected by variance threshold and those selected by mutual information. The top 5,000 features by variance threshold and mutual information shared 3,958 features and the top 10,000 features from both shared 8,598 features. On the other hand, there were only 344 preserved features present in the top 10,000 and 2,278 in the top 20,000 lists produced by mutual information and PCC. The top 20 features ranked by mutual information and variance threshold are provided in Table 3.

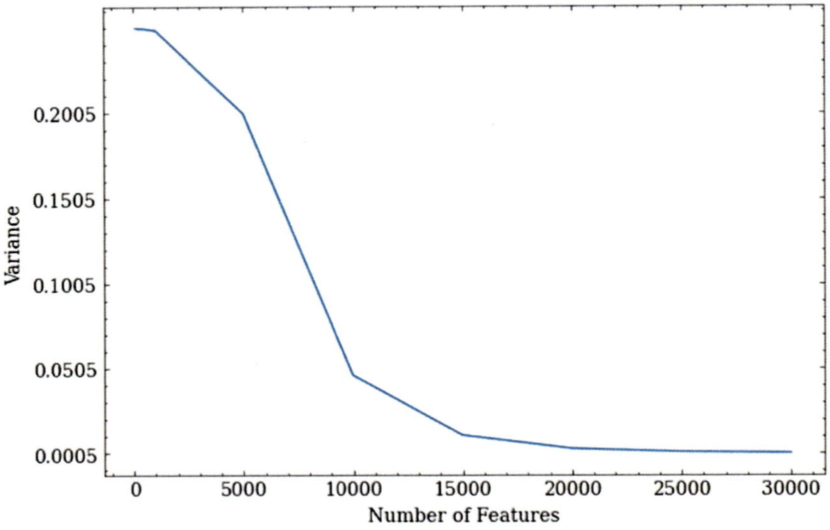

Fig. 2. Number of features selected by variance thresholds

Figure 2 shows the variance values for the feature set sizes of 100 to 30,000. A substantial drop in variance after 5,000 features can be seen from the plot. The fall in variance signifies that many of the features do not greatly impact the accuracy of the models. Therefore, the accuracy rates for mutual information and variance threshold increased considerably only up to 10,000 features, coinciding with the plot in Fig. 2, where there is a marked decrease in variance after 10,000 features. We further investigated the best threshold value between 0.05 and 0.25 to identify the variance threshold that produces the best accuracy metrics. The threshold of 0.15 achieved the highest accuracy with 6,443 features.

Table 3. Top 20 features ranked by mutual information and variance threshold

Rank	Mutual information	Variance threshold
1	Landroid/content/pm/PackageInstaller$SessionInfo;->getAppPackageName	Landroid/content/res/Resources$Theme;->applyStyle
2	Landroid/content/pm/PackageInstaller;->getAllSessions	Landroid/app/Notification$Builder;->setCustomHeadsUpContentView
3	Landroid/os/DeadObjectException;-><init>	Landroid/view/accessibility/AccessibilityNodeInfo;->setViewIdResourceName
4	Landroid/os/UserManager;->getApplicationRestrictions	Landroid/content/Context;->getObbDirs
5	Landroid/content/pm/PackageManager;->getPackageInstaller	Landroid/media/MediaPlayer;->setOnErrorListener
6	Landroid/os/Binder;->restoreCallingIdentity	Landroid/graphics/drawable/LayerDrawable;->setId
7	Landroid/os/Binder;->clearCallingIdentity	Landroid/content/pm/PackageManager;->queryIntentActivityOptions
8	Landroid/os/IInterface;->asBinder	Landroid/os/Parcel;->writeValue
9	Landroid/os/Parcel;->dataPosition	Landroid/widget/ProgressBar;->setIndeterminate
10	Landroid/app/ActivityManager;->getMyMemoryState	Landroid/widget/ListView;->setOnKeyListener
11	Landroid/content/ServiceConnection;->onServiceConnected	Landroid/text/Editable;->length
12	Landroid/content/ServiceConnection;->onServiceDisconnected	Landroid/view/accessibility/AccessibilityNodeInfo;->setContentInvalid
13	Landroid/app/AppOpsManager;->checkPackage	Ljava/io/FileNotFoundException;->printStackTrace
14	Landroid/content/pm/PackageManager;->isInstantApp	Landroid/content/Context;->getObbDir
15	Landroid/os/PowerManager;->isInteractive	Landroid/app/Notification$Builder;->setCustomContentView
16	Landroid/os/Parcel;->dataSize	Landroid/view/MenuItem;->getIcon
17	Ljava/util/logging/Logger;->logp	Landroid/database/Cursor;->getExtras
18	Ljava/lang/Character;->isSurrogatePair	Landroid/hardware/display/DisplayManager;->getDisplays
19	Ljava/util/TreeMap;->descendingMap	Landroid/app/Notification$Builder;->setCustomBigContentView
20	Landroid/view/ViewGroup;->startViewTransition	Landroid/os/ResultReceiver;-><init>

4 Conclusion

As the popularity of the Android OS has increased in recent years, it has become a major target for malware developers. This can result in personal data becoming vulnerable and calls for developing robust anti-malware techniques. Research shows machine learning techniques work well in identifying new and old malware.

API calls is one of the most used features in Android malware detection using machine learning. However, due to the number of APIs provided by the Android SDK, the number of API calls used by applications can become overwhelming from a machine learning perspective. In our dataset of 40,000 applications, 134,207 different API calls were used. In this work, we analysed how different ML models, namely SVM, decision tree, random forest, Naïve Bayes and AdaBoost, perform with the API feature set and how the API feature set can be reduced for more practical use with three feature selection algorithms, mutual information, variance threshold and Pearson Correlation Co-efficient. The results show that random forest classifiers perform the best when used with an API calls feature set, and we can reduce the number of features by 95.2% to achieve better detection accuracy than the complete API call feature set.

References

1. Turner, A.: How many smartphones are in the world? (2021). https://www.bankmycell.com/blog/how-many-phones-are-in-the-world. Accessed 31 July 2021
2. StatCounter: Mobile operating system market share worldwide (2021). http://gs.statcounter.com/os-market-share/mobile/worldwide. Accessed 16 Apr 2021
3. Google play store. https://play.google.com/store
4. Osborne, C.: Joker billing fraud malware found in google play store (2021). https://www.zdnet.com/article/joker-billing-fraud-malware-found-in-google-play-store/. Accessed 31 July 2021
5. Yu, B., Fang, Y., Yang, Q., Tang, Y., Liu, L.: A survey of malware behavior description and analysis. Front. Inf. Technol. Electron. Eng. **19**(5), 583–603 (2018). https://doi.org/10.1631/FITEE.1601745
6. Peiravian, N., Zhu, X.: Machine learning for android malware detection using permission and API calls. In: 2013 IEEE 25th International Conference on Tools with Artificial Intelligence, pp. 300–305. IEEE, November 2013. http://ieeexplore.ieee.org/document/6735264/
7. Arp, D., Spreitzenbarth, M., Hübner, M., Gascon, H., Rieck, K.: DREBIN: effective and explainable detection of android malware in your pocket. In: Network and Distributed System Security Symposium (NDSS), no. August (2014)
8. Ma, Z., Ge, H., Liu, Y., Zhao, M., Ma, J.: A combination method for android malware detection based on control flow graphs and machine learning algorithms. IEEE Access **7**(c), 21 235–21 245 (2019)
9. Jung, J., et al.: Android malware detection based on useful API calls and machine learning. In: Proceedings - 2018 1st IEEE International Conference on Artificial Intelligence and Knowledge Engineering, AIKE 2018, pp. 175–178 (2018)
10. Afonso, V.M., de Amorim, M.F., Grégio, A.R.A., Junquera, G.B., de Geus, P.L.: Identifying Android malware using dynamically obtained features. J. Comput. Virol. Hack. Tech. **11**(1), 9–17 (2014). https://doi.org/10.1007/s11416-014-0226-7

11. Xiao, X., Zhang, S., Mercaldo, F., Hu, G., Sangaiah, A.K.: Android malware detection based on system call sequences and LSTM. Multimed. Tools Appl. **78**(4), 3979–3999 (2017). https://doi.org/10.1007/s11042-017-5104-0
12. Saracino, A., Sgandurra, D., Dini, G., Martinelli, F.: MADAM: effective and efficient behavior-based android malware detection and prevention. IEEE Trans. Dependable Secure Comput. **15**(1), 83–97 (2018)
13. Package index (2021). https://developer.android.com/reference/packages. Accessed 31 July 2021
14. R. Connor Tumbleson: Apktool (2019). https://ibotpeaches.github.io/Apktool/
15. Virusshare. https://virusshare.com
16. App downloads for android. https://en.uptodown.com/android. Accessed 31 July 2021
17. Apkmirror. https://www.apkmirror.com/. Accessed 31 July 2021
18. F-droid - free and open source android app repository. https://www.f-droid.org/. Accessed 31 July 2021
19. Virustotal. https://www.virustotal.com
20. Guyon, I., Elisseeff, A.: An introduction to variable and feature selection. J. Mach. Learn. Res. **3**(null), 1157–1182 (2003)

Intrusion Detection for CAN Using Deep Learning Techniques

Rawan Suwwan(✉), Seba Alkafri, Lotf Elsadek, Khaled Afifi, Imran Zualkernan, and Fadi Aloul

Department of Computer Science and Engineering, American University of Sharjah, Sharjah, UAE
{g00075916,g00073729,b00075151,b00073944,izualkernan, faloul}@aus.edu

Abstract. With the advent of Internet of Vehicles (IoV), cars and commercial vehicles represent a convenient attack surface for cyber attacks. Many automobiles use the Controller Area Network (CAN) bus for internal communication. CAN is known to be susceptible to various types of cyber attacks. One constraint on intrusion detection systems (IDS) for CAN is that they need to be efficient due to lack of resources and the high traffic on a typical CAN network. This paper presents an implementation of simple 1D Convolutional Neural Network (CNN), Long Short Term (LSTM) and Gated Recurrent Units (GRU) networks on a recent attack data set for CAN. All models thus developed outperformed the existing state-of-art and achieve an almost perfect F1-Score of 1.0.

Keywords: CAN attacks · Cybersecurity · Deep learning · GRU · LSTM · CNN

1 Introduction

As modern automobiles are increasingly digital, cyber attacks on board networks like the Controller Area Network (CAN) bus pose potentially fatal consequences. With the advent of 5G, the Internet of Vehicles (IoV) is fast becoming a reality [1]. Therefore, connected cars will be at an increasing risk of being attacked for malicious purposes. This paper explores the implementation of an intrusion detection system (IDS) that detects CAN attacks. An IDS is a software or hardware security tool that detects attacks that cannot be prevented by other security mechanisms and responds to mitigate the effects of the attack. CAN is a standardized message-based protocol widely used in vehicles for communication. CAN is a network bus that connects all the different components or ECUs (Engine Control Unit) in the car. In an automotive CAN bus system, ECUs may include the engine control unit, airbags, audio system or other components. A modern car can have up to 70 ECUs, where each of them transmits information that needs to be shared with other parts of the network [2]. CAN is currently the standard in today's vehicles as per the CAN FD standards (ISO 11898-1 and ISO 11898-2). Figure 1 shows the standard components of a CAN data frame. CAN data can be vulnerable to malicious monitoring as they are transmitted via broadcast. Furthermore, encryption is not used which can lead to the sniffing and hacking of the data.

© The Author(s), under exclusive license to Springer Nature Switzerland AG 2022
H. Ragab Hassen and H. Batatia (Eds.): ACS 2021, LNNS 378, pp. 13–19, 2022.
https://doi.org/10.1007/978-3-030-95918-0_2

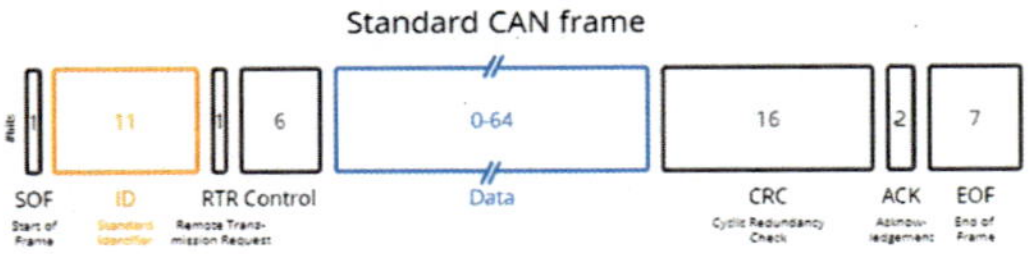

Fig. 1. A standard CAN frame.

2 Previous Work

Table 1 shows a summary of previous work in intrusion detection for CAN networks. As Table 1 shows, three common data sets (i.e., [2–4]) in addition to a number of custom data set have been used. This makes it difficult to compare results across studies. Data from a variety of vehicles including Kia, Hyundai, Honda, Dodge, Suzuki, etc. has been used. As Table 1 shows, a variety of techniques including signature based (e.g., [5, 6]), traditional machine learning (e.g., [7–10]), deep learning (e.g., [11–13]), and unsupervised learning (e.g., [14, 15]) have been explored. The generally considered attacks include Denial of Service (DoS), Impersonation, Fuzzing, and Spoofing of Gear or RPM packets. In addition, other attacks like Replay, Injection, Camouflage have also been explored. Finally, as Table 1 shows, most techniques have yielded impressive results. In this paper we explore the most recent commonly used data set [4] because it is publicly available and hence allows for direct comparison to other work.

3 Dataset and Feature Engineering

Table 2 shows a breakdown of the classes in the dataset [4]. This dataset was also collected for different states for the car (stationary vs. driving). This paper used the data from the driving round, which comprised of 2,000,733 data points. Each data point includes a Timestamp (logging time), Arbitration_ID (CAN identifier), DLC (data length code), Data (CAN data field), Class (Normal or Attack), and SubClass (attack type) of each CAN message. For example, one datapoint may look like `16.05236,130,8,14 80 10 80 00 00 0A 73`. The ID (e.g., `130`) and the DLC (e.g., `8`) were discarded. The Data field contains the actual packet data (e.g., `14 80 10 80 00 00 0A 73)` from the CAN frame. Since data length was arbitrary, the ending bits were padded with 00 in case the data length was shorter than 8 bytes. The data was scaled and since the data was unbalanced, Synthetic Minority Oversampling Technique (SMOTE) was used to balance the data. Each of the datapoints was labelled with one of the Sub-class attack types as shown in Table 2. This resulted in a multi-class classification problem.

Table 1. Summary of previous work

Ref	Attacks	Dataset	Vehicle	Techniques	Results
Lee et al. [5] (2017)	DoS, Imp., Fuzzy	[2]	Kia Soul	Signature based	Could detect attacks based on time
Moulahi et al. [7] (2021)	DoS, Imp., Fuzzy	[2]	Kia Soul	RF, DT, SVM, DTD	Accuracy 98.1%–98.5%
Javed et al. [11] (2021)	DoS, Imp., Fuzzy	[2]	Kia Soul	CNN + Attention-GRU	F1-Score 93.9–94.38
Seol et al. [16] (2018)	DoS, Fuzzy, Gear, RPM	[3]	Hyundai's YF Sonata	GAN	Accuracy 99.6%–99.9%
Song et al. [12] (2020)	Gear, RPM	[3]	Hyundai's YF Sonata	RESNET + LSTM	Accuracy 91%
Amato et al. [13] (2021)	Dos, Fuzzy, Gear, RPM	[3]	Hyundai's YF Sonata	DNN	Accuracy 98%–100%
Mehedi et al. [17] (2021)	Dos, Fuzzy, Gear, RPM	[4]	Hyundai Avante CN7	1D CNN	Accuracy 97.8–98.1% F1-Score 0.92–0.95
Hanselmann et al. [15] (2020)	Plateau, change, Playback, Flooding, Suppress	[18]	Unknown	LSTM + AutoEnco,	Accuracy 99.1–99.2%
Omid et al. [8] (2019)	DoS, Fuzzy	Custom	Dodge RAM Pickup	OCSVM-MBA	Accuracy 95.5%–97%
Zhou et al. [19] (2019)	Abnormal	Custom	Unknown	Siamese Triplet DNN	Accuracy 83%
Qin et al. [20] (2021)	Replay, Temper	Custom	Unknown	LSTM	F1-Score 85%
Delwar et al.[21] (2021)	DoS, Fuzzy, Spoofing	Custom	Toyota, Subaru, Suzuki	1D CNN	Accuracy 99.8%
Xun et al. [22] (2021)	Abnormal	Custom	Luxgen U5, Buick Regal	Deep SVDD	Accuracy 98.5%
Li et al. [9] (2021)	Abnormal	Custom	Luxgen U5	M-SVDD, G-SVDD	Accuracy 98.37%–99.53%
He et al. [10] (2021)	Injection, Camouflage, Suspension, Tempering,	Custom	Jeep and Unknown	LightGBM	F1-Score 90.49–100
Jin et al. [6] (2021)	Drop, Replay, Tempering	Custom	Unknown	Signature-based	Accuracy: 66%–100%
Leslie [14] (2021)	Abnormal	Custom	Unknown	Ensemble Clustering	F1-Score 100

Table 2. Class breakdown of dataset

Sub-Class	Description	
	Definition	Type
Normal	Normal traffic in CAN bus	Normal
Flooding (DoS)	Flooding attack aims to fill the CAN bus segment with a massive number of traffic messages so that the network bus is congested and hence prevents the targeted service traffic to come through	Attack
Spoofing	CAN messages are injected to control certain desired functions as the source destination is spoofed	Attack
Replay	Replay attack is to extract normal traffic at a specific time and replay (inject) it into the CAN bus	Attack
Fuzzing	Random messages are injected to cause unexpected behavior of the vehicle	Attack

4 Neural Network Architectures

As Table 1 shows, Mehedi et al. [17] used a 1DCNN to achieve a an F1-Score of 0.92 to 0.95 on this data set. However, as Table 1 shows, LSTMs (e.g., [12, 15, 20]) and GRUs (e.g., [11]) have been used successful for ID as well. In addition, time difference between the blocks arriving seems to be a useful feature for intrusion detection (e.g., [5, 6]). Therefore, we considered the time difference as well as the padded data as inputs to 1DCNN, LSTM, Bidirectional LSTM and a GRU model. The LSTM, Bidirectional LSTM, and GRU used a Dense()-Dropout(0.35)-Dense(4) network using the SGD optimizer, learning rate of 0.1, and cross-entropy loss. The 1DCNN used a Conv1D(8)-Dropout(0.25)-MaxPooling1D-Flatten-Dense(ReLU)-Dense(4) network. Each of the above model was trained using a 60/20/20 training/validation/testing split for a total of 30 epochs each. No overfitting was observed based on the loss curves for any of the models.

5 Results

Figure 2 shows the performance metrics for the best models of each type. As the Figure shows, all four models were able to achieve high F1-scores of at least 0.99 or more and hence outperforming Mehedi et al. [17]. This can clearly be attributed to the inclusion of the time difference data in addition to the packet data.

	Precision	Recall	F1-Score	Support
Normal	0.99	0.99	0.99	13384
Flooding	1.00	1.00	1.00	11337
Fuzzing	1.00	1.00	1.00	11618
Replay	0.99	0.99	0.99	11670
Spoofing	1.00	1.00	1.00	11959
Accuracy			1.00	59968
Macro Average	1.00	1.00	1.00	59968
Weighted Average	1.00	1.00	1.00	59968

(a) LSTM

	Precision	Recall	F1-Score	Support
Normal	1.00	1.00	1.00	13384
Flooding	1.00	1.00	1.00	11337
Fuzzing	1.00	1.00	1.00	11618
Replay	1.00	1.00	1.00	11670
Spoofing	1.00	1.00	1.00	11959
Accuracy			1.00	59968
Macro Average	1.00	1.00	1.00	59968
Weighted Average	1.00	1.00	1.00	59968

(b) GRU

	Precision	Recall	F1-Score	Support
Normal	1.00	1.00	1.00	13384
Flooding	1.00	1.00	1.00	11337
Fuzzing	1.00	1.00	1.00	11618
Replay	1.00	1.00	1.00	11670
Spoofing	1.00	1.00	1.00	11959
Accuracy			1.00	59968
Macro Average	1.00	1.00	1.00	59968
Weighted Average	1.00	1.00	1.00	59968

(c) 1D CNN

	Precision	Recall	F1-Score	Support
Normal	0.98	1.00	0.99	13384
Flooding	1.00	1.00	1.00	11337
Fuzzing	1.00	1.00	1.00	11618
Replay	0.99	0.98	0.99	11670
Spoofing	1.00	1.00	1.00	11959
Accuracy			0.99	59968
Macro Average	0.99	0.99	0.99	59968
Weighted Average	0.99	0.99	0.99	59968

(d) BiDirectional LSTM

Fig. 2. Performance metrics for the best models

Table 3 summarizes the results of K-fold testing confirming that the 1DCNN had the best mean macro F1-Score of 0.9997 with a very small standard deviation of 9.2223e-5 showing that this is a robust model in addition to being accurate. Figure 3 shows the results of 10-fold testing showing that 1DCNN seemed to have performed the best overall with the greatest number of high F1-Score models.

Table 3. K-fold testing results (K = 10)

Method	Mean Macro F1-Score	Standard Deviation
LSTM	0.9944	0.00230
GRU	0.9993	0.00035
1DCNN	0.9997	9.2223e-5
BiDirectional LSTM	0.9944	0.002304

An analysis of the confusion matrices showed that Replay and Normal class were the most misclassified across the various methods.

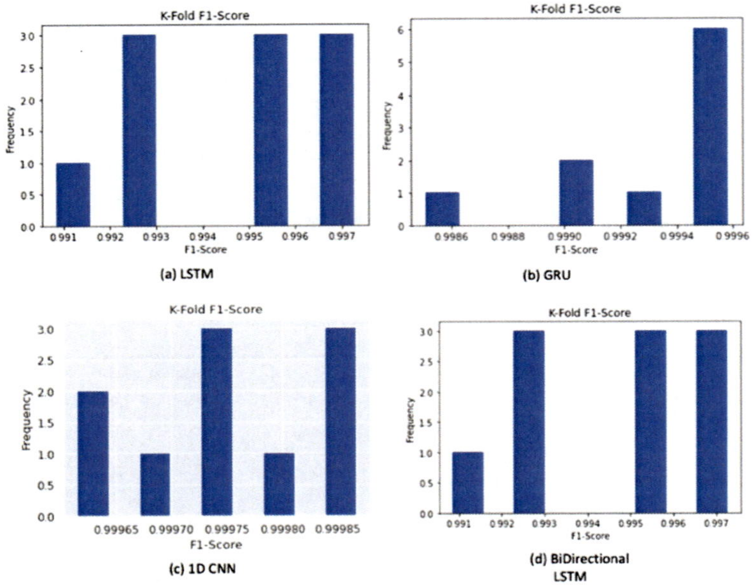

Fig. 3. 10-Fold Macro-F1 metrics

6 Conclusion

While many ID models have been proposed for CAN networks, this paper has presented the best state-of-the-art results by using very conventional and small neural network models.

References

1. Martínez, I.: The 5g car. In: Martínez, I. (ed.) The Future of the Automotive Industry The Disruptive Forces of AI, Data Analytics, and Digitization, pp. 45–62. Apress, Berkeley, CA (2021). https://doi.org/10.1007/978-1-4842-7026-4_3
2. [HIDE]CAN-intrusion-dataset (OTIDS). https://ocslab.hksecurity.net/Dataset/CAN-intrusion-dataset. Accessed 07 Aug 2021
3. Car-Hacking Dataset. https://ocslab.hksecurity.net/Datasets/CAN-intrusion-dataset. Accessed 07 Aug 2021
4. Kim, H.K. : Car Hacking: Attack & Defense Challenge 2020 Dataset. IEEE, 03 Feb 2021. https://ieee-dataport.org/open-access/car-hacking-attack-defense-challenge-2020-dataset Accessed 05 Aug 2021
5. Lee, H., Jeong, S.H., Kim, H.K.: OTIDS: a novel intrusion detection system for in-vehicle network by using remote frame. In: 2017 15th Annual Conference on Privacy, Security and Trust (PST), pp. 57–5709 (August 2017). https://doi.org/10.1109/PST.2017.00017
6. Jin, S., Chung, J.-G., Xu, Y.: Signature-based intrusion detection system (IDS) for in-vehicle can bus network. In: 2021 IEEE International Symposium on Circuits and Systems (ISCAS), pp. 1–5 (May 2021). https://doi.org/10.1109/ISCAS51556.2021.9401087

7. Moulahi, T., Zidi, S., Alabdulatif, A., Atiquzzaman, M.: Comparative performance evaluation of intrusion detection based on machine learning in in-vehicle controller area network bus. IEEE Access **9**, 99595–99605 (2021). https://doi.org/10.1109/ACCESS.2021.3095962
8. Avatefipour, O., et al.: An intelligent secured framework for cyberattack detection in electric vehicles' CAN bus using machine learning. IEEE Access **7**, 127580–127592 (2019). https://doi.org/10.1109/ACCESS.2019.2937576
9. Li, X., et al.: CAN bus messages abnormal detection using improved SVDD in internet of vehicle. IEEE Internet Things J. 1 (2021). https://doi.org/10.1109/JIOT.2021.3098221
10. He, X., Yang, Z., Huang, Y.: A Vehicle intrusion detection system based on time interval and data fields. In: Sun, X., Zhang, X., Xia, Z., Bertino, E. (eds.) ICAIS 2021. LNCS, vol. 12737, pp. 538–549. Springer, Cham (2021). https://doi.org/10.1007/978-3-030-78612-0_43
11. Javed, A.R., Ur Rehman, S., Khan, M.U., Alazab, M., Reddy, T.G.: CANintelliIDS: detecting in-vehicle intrusion attacks on a controller area network using CNN and attention-based GRU. IEEE Trans. Netw. Sci. Eng. **8**(2), 1456–1466 (2021). https://doi.org/10.1109/TNSE.2021.3059881
12. Song, J., Li, F., Li, R., Li, Y., Zhou, Q., Zhang, J.: Research on CAN bus anomaly detection based on LSTM AndResNet. J. Phys. Conf. Ser. **1757**(1), 012044 (2021). https://doi.org/10.1088/1742-6596/1757/1/012044
13. Amato, F., Coppolino, L., Mercaldo, F., Moscato, F., Nardone, R., Santone, A.: CAN-Bus attack detection with deep learning. IEEE Trans. Intell. Transp. Syst. 1–10 (2021). https://doi.org/10.1109/TITS.2020.3046974
14. Leslie, N.: An unsupervised learning approach for in-vehicle network intrusion detection. In: 2021 55th Annual Conference on Information Sciences and Systems (CISS), pp. 1–4 (March 2021). https://doi.org/10.1109/CISS50987.2021.9400233
15. Hanselmann, M., Strauss, T., Dormann, K., Ulmer, H.: CANet: an unsupervised intrusion detection system for high dimensional CAN bus data. IEEE Access **8**, 58194–58205 (2020). https://doi.org/10.1109/ACCESS.2020.2982544
16. Seo, E., Song, H.M., Kim, H.K.: GIDS: gan based intrusion detection system for in-vehicle network. In: 2018 16th Annual Conference on Privacy, Security and Trust (PST), pp. 1–6 (August 2018). https://doi.org/10.1109/PST.2018.8514157
17. Mehedi, S.T., Anwar, A., Rahman, Z., Ahmed, K.: Deep transfer learning based intrusion detection system for electric vehicular networks. Sensors **21**(14) (2021). Art. no. 14, https://doi.org/10.3390/s21144736
18. SynCAN Dataset. ETAS (2021). https://github.com/etas/SynCAN. Accessed 07 Aug 2021
19. Applied Sciences | Free Full-Text | Anomaly Detection of CAN Bus Messages Using a Deep Neural Network for Autonomous Vehicles | HTML. https://www.mdpi.com/2076-3417/9/15/3174/htm. Accessed 05 Aug 2021
20. Qin, H., Yan, M., Ji, H.: Application of controller area network (CAN) bus anomaly detection based on time series prediction. Veh. Commun. **27**, 100291 (2021). https://doi.org/10.1016/j.vehcom.2020.100291
21. Delwar Hossain, M., Inoue, H., Ochiai, H., Fall, D., Kadobayashi, Y.: An Effective in-vehicle CAN bus intrusion detection system using CNN deep learning approach. In: GLOBECOM 2020 - 2020 IEEE Global Communications Conference, pp. 1–6 (December 2020). https://doi.org/10.1109/GLOBECOM42002.2020.9322395
22. Xun, Y., Zhao, Y., Liu, J.: VehicleEIDS: a novel external intrusion detection system based on vehicle voltage signals. IEEE Internet Things J. 1 (2021). https://doi.org/10.1109/JIOT.2021.3090397

Check for updates

A Comparative Study of Machine Learning Binary Classification Methods for Botnet Detection

Nadim Elsakaan(✉) and Kamal Amroun

LIMED – Faculty of Exact Sciences, University of Bejaia, Bejaia, Algeria
{nadim.elsakaan,kamal.amroun}@univ-bejaia.dz

Abstract. Internet of things is one of the most important numerical revolution, it offers a lot of opportunities by interconnecting all everyday objects, unfortunately due to their nature they are weaker face to hacker's attacks. Botnets are one of the biggest threat against actual internet because of their capacity to realize distributed denial of services (DDoS) attacks at any time, they will be more devastating by taking advantage of edge-compute capacities of connected objects which become a phenomenal treatment unit when combined. Lot of works try to anticipate the propagation of botnets brought by the IoT development and to propose solutions based on machine and deep learning to face them. In this work we explore existing datasets in classic internet, proceed to features categorization on the CSE-CIC-IDS2018 one, then we train several machine learning models with each of the proposed subsets, evaluate them and try to project the same procedure on a dataset dedicated to internet of things (Botnet-IoT).

1 Introduction

The internet of things (IoT) or the third generation of internet is a global network interconnecting devices which are everyday objects like cars, watches, televisions or anything else enhanced with some characteristics: (i) computation and calculation capacities, (ii) communication on the network with a unique global identifier and (iii) a capacity to interact with the environment to capture parameters such as temperature (sensors) or to in influence it like lifting curtains (actuator). All this is made possible by the work of many researchers who have allowed a high interoperability between a set of key technologies including wireless sensor networks (WSN), cloud and edge computing, peer-to-peer networks with several communication patterns like machine-to-machine or human-to-machine and so on.

Convergence towards the Internet of Things brings up several security challenges. Indeed, the IoT being an interconnection of existing technologies it is often also qualified as an interconnection of adjacent threats [7]. In addition to classic vulnerabilities, problems related to the nature of connected objects appear (if they are not properly configured and secured), we can cite: weak encryption

© The Author(s), under exclusive license to Springer Nature Switzerland AG 2022
H. Ragab Hassen and H. Batatia (Eds.): ACS 2021, LNNS 378, pp. 20–34, 2022.
https://doi.org/10.1007/978-3-030-95918-0_3

due to the computational limited capabilities, low-energy level since objects are powered by batteries, continuous identification which leads to infringement of privacy.

Distributed denial of service (DDoS) attacks are one of the biggest threats to networks, these kind of attacks are generally based on botnets (bot networks), a bot is a malicious software that runs on the victim's machine without his approval, then a botnet can be defined as a network of compromised computers that have been infected by bots. The danger of botnets lies in their ability to launch DDoS attacks at any time.

To face this security threat, lot of solutions were proposed in the literature to enhance intrusion detection systems (IDS) taking advantage of the most powerfull data analysis and machine learning algorithms. It's worth pointing that IDS monitor networks (network based systems) and computers (host based systems) in order to detect intrusion attempts. There are three main class of approaches used against botnet: (i) detection of infected machines [15], (ii) detection of suspicious network traffic [4] and (iii) attack prevention which provides interesting tools allowing to prevent the moment when the botnet are going to execute an attack [1].

The machine learning algorithms explore data and search attack sequences by identifying correlation between independent variables (features) and the target (label), then learn from it patterns in order to prevent and detect later suspicious behaviour in the network activity. Training this kind of algorithms requires a huge volume of data with various features in order to select those giving the maximum level of information to detect and prevent an attack. Lot of datasets exist to meet these needs, in the Subsect. 2.2 we describe some of them which are the most used by searchers and propose a categorization of them into three main classes: (i) flow information based datasets, (ii) packets information based datasets or (iii) application based datasets.

In this paper, we explore two of the most commonly used datasets, our aim is to propose in Subsect. 2.2 a classification of datasets and then in Sect. 3 a categorization of features then we carry out some experiments to demonstrate the most relevant ones for the botnet detection. Given the lack of cybersecurity dataset in the context of the Internet of Things we propose to start analyzing classic network datasets and then to project our conclusions on the IoT-botnet dataset which is, as best as we know, the only cybersecurity dataset dedicated to botnet in the internet of things. Also, in Sect. 3 we propose a methodology to build machine learning based intrusion detection system for botnets. To validate our study we train several binary classification models and present the corresponding results which seem to be interesting in the Sect. 4.

The rest of this article is structured as follows: Sect. 2 presents the related works that we considered to be the most consistent. Section 3 is devoted to our contribution. Section 4 is dedicated for performance evaluation and discussion of the experiments results, this is an ongoing research paper and presented results are intermediary ones. We finish with the Sect. 5 which concludes our work and introduces some interesting perspectives.

2 Background

In this section, we give an overview of the related works on data science applications for cybersecurity and briefly present some works dedicated to botnet detection.

2.1 Intrusion Detection System (IDS)

An IDS is a security system used to monitor networks (NB-IDS) and computers (HB-IDS) in order to detect intrusion attempts [11]. It is a mechanism used to detect attacks against a network by analyzing the network traffic and notifying the administrator [3]. There are two main approaches for Intrusion Detection:

- Signature Based: The tool attempts to match the current network behavior with saved attack patterns, so a signature of each attack needs to be saved before the system deployment.
- Anomaly based: The tool make use of machine learning models and attempts to detect unknown suspicious behaviour and alerts administrators.

2.1.1 Machine Learning Based IDS

Machine learning based classification algorithms find native application in cybersecurity particularly in the area of intrusion detection, many approaches exist in the literature to deal with this problem and especially with DDoS attacks.

Mohamed Amine F et al. proposed a comparative study of deep learning approaches for intrusion detection, they trained several deep learning models like recurrent and convolutional neural networks, deep neural networks, restricted and deep Boltzman machines, deep auto-encoders for binary and multi-class classification with two datasets, namely: CSE-CIC-IDS2018 and the Bot-IoT datasets. The best accuracy for botnet detection based on deep neural network is 98.221% given by using a learning rate of 0.5 and one hundred hidden nodes for the IoT-Botnet dataset against 97.281% for the CSE-CIC-IDS2018 dataset with same configuration. Nevertheless, we note the absence of a phase of features selection this lead to a longer execution time and a denser network [8].

By using a simple features classification of the UNSW-NB15 dataset: (i) basic, (ii) content, (iii) traffic characteristics based: on time, on source address, on destination address, authors of [16] trained recurrent neural network based on LSTM architecture and multi-layer neural networks (MLP). They obtained the best accuracy of 98.8% with MLP for the second group with the linear rectifier, Adam optimizer and a custom calculation rule for hidden layer nodes.

These and a lot of other works have taken advantage of machine learning approaches to improve intrusion detection by rendering IDS more reactive against new attack patterns, in their overview on cybersecurity based on data science, Iqbal H.Sarker et al. [22] have summarized numerous solutions which were proposed to this end.

2.1.2 Botnet Detection and Prevention Approaches

As previously said, due to their capacity to execute a DDoS attack at any time with a huge volume of packets, botnet stay one of the most critical threat cybersecurity has to face in the future generation of internet. Indeed, the increase of number of connected objects, with limited security capacities allow attackers to expand the number of their zombie machines. Some searchers try to model the development and the propagation of botnet in the internet of things. In [26] authors tried to propose a dynamic botnet propagation model based on susceptible-exposed-infected-recovered (SEIR) epidemic model taking in account social characteristics of users.

Botnet life-cycle can be depicted in 5 main phases: malware development, malware injection, takeover of the infected machines, information and resources exploitation, malware maintaining in the machines [12]. Intrusion detection techniques intervene at different levels of this cycle, some solutions focus on detection of infected machines when other analyze network traffic to detect connection between them and command & control servers. An example is the solution of R. Vinayakumar et al. [25] who trained neural networks based on many architectures in order to analyze logs of DNS services and looking for Domain Generation Algorithms (DGA) generated names which are used by c&c servers in order to connect to infected machines.

On the other hand, some approaches take advantage of edge computing capacities of object, like Georgios Spathoulas et al. [23] who use it to create a collaborative block-chain based detector of DDoS. More globally, [13] gives a classification of network forensic methods for investigating botnets: honeypots, network flow analysis, deep packet analysis, attack recognition and visualisation of network traffic. Another model is proposed in [5], it extracts local features from devices CPU like usage rate and number of process then realize binary classification.

2.2 Datasets

There are many cybersecurity datasets commonly used by searcher to train several kinds of machine learning algorithms, they were created thanks to different tools, like network simulators and traffic generators. According to the context of datasets related to the network in which they were generated and their objectives, some classifications exist to order them. An example is given by Mohamed Amine F et al. and categorize datasets into seven classes: network traffic-based dataset, electrical network-based dataset, internet traffic-based dataset, virtual private network-based dataset, android apps-based dataset, IoT traffic-based dataset, and internet-connected devices-based dataset [8].

For our work, we will propose a simpler and more intuitive classification, which divides the datasets according to the contents into three main categories:

1. Flow based datasets: This is a category of datasets composed of information about flows and statistics on traffic like connections duration, number of packets in a tcp sub-flow and so on.
   - KDD cup 1999 this dataset is issued from the 1998 DARPA intrusion detection evaluation program, it contains four gigabytes of compressed

binary TCP dump as raw training data including four main categories of attacks: Denial of Service (DoS), unauthorized access from a remote machine (R2L) or to a local machine (U2R), and probing like port scanning [24].
- CSE-CIC-IDS2018 this dataset includes standard flow information of a network composed of 420 machines, 30 servers and malicious traffic generated by an attack-network of 50 machines. Attacks cover brute-force, DoS, DDoS, infiltration and botnet attack [10].
- IoT_Botnet_Dataset this dataset was created by designing a realistic network environment in the Cyber Range Lab of UNSW Canberra, the captured pcap files of 69.3 gigabytes includes 72 millions of records from which flow traffic information were extracted into 16 gigabytes csv files covering DDoS, DoS, operating system and service scan, key-logging and ex-filtration attacks [14].

2. Packets based datasets: datasets of this category are composed of network sniffing files like pcap ones, they include basic information like ip addresses, ports and flags.
   - ISOT_Botnet_DataSet_2010 this dataset is combination of existing public datasets containing begnin and malicious traffic, it includes 161 millions of packets captured over four days on a network infected with three botnet virus: Waledac, Zeus and Storm [20].
   - CICDataset-ISCX-Bot-2014 this dataset composed of training and test datasets respectively with size of 5.3 GB and 8.5 GB containing approximately 44% of malicious flows, this dataset include 7 botnets: Ners, Rbot, Virut, NSIS, SMTP SPAM, Zeus(peer-to-peer -P2P- and command and control-C&C-) [9].
   - UNSW-NB15 this dataset is created by IXIA PerfectStorm tool in the Cyber Range Lab of the Australian Center for cybersecurity, it contains 175 thousands of records in training set and 82 thousands in the test one. It covers 9 categories of attacks, namely: fuzzers, analysis, back-doors, DoS, exploits, generic, reconnaissance, shell-code and worms [18].
   - CTU-13-Dataset this dataset was captured in the CTU university, Czech Republic, its main goal was to capture large traffic with real botnet and normal flows, the following bots were used: Neris, Rbot, Virut, Menti, Sogou, Murlo, NSIS [21].
   - CTU-23-IoT-Dataset this dataset of network traffic from IoT objects mixed with 20 malware captured during execution. Among the contained malicious netflows we found botnet ones based on Mirai, IRCBot, Okiru, Torii and Kenjiro [19].
   - NSL-KDD this dataset was proposed to solve some of the inherent problems of the KDD99, it removes redundant records and the number of selected records is more representative to allow high variation in accuracy of trained models [24].
3. application based datasets: this category encompasses datasets with application samples infected by virus.

- AndroidBotnets_Dataset is dedicated to evaluation of android botnets, it contains 1929 samples of application spawning a period of 2010 to 2014, these samples represent 14 botnets families including: AnserverBot, Bmaster, DroidDream, Geinimi, MisoSMS, NickySpy, Not Compatible, PJapps, Pletor, RootSmart, Sandroid, TigerBot, Wroba, Zitmo [2].
- CICMalDroid_2020 this is one of the most recent (samples until 2018) android malware dataset, it is big with approximately 17 thousands of android samples, diverse and comprehensive thanks to a large variety of malware: ad-wares, banking malware, sms malware and mobile risk-wares [17].

On the other hand, some works proposed a framework to allow bench-marking existing datasets and generating new ones based on real situation pr artificial intelligence techniques in order to improve intrusion detection rate. Several suggested approaches can be cited: simulation, emulation, physical network building. The resulting dataset can be evaluated mainly by checking: exhaustiveness of network configuration, the coherence of traffic generation method, dataset labels, used protocols, diversity of attacks [10].

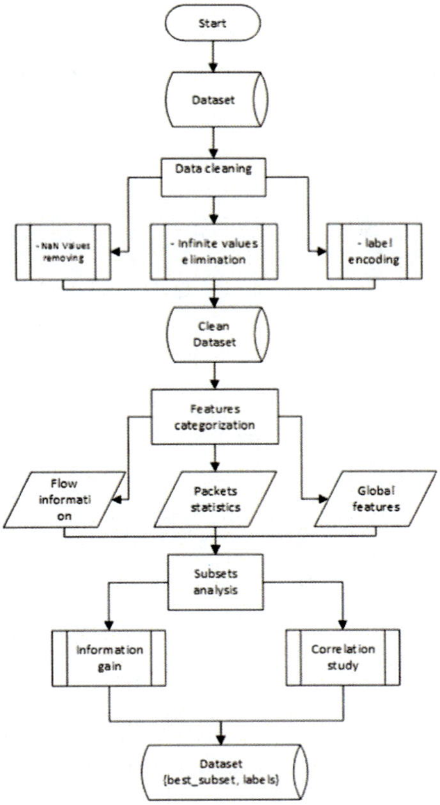

Fig. 1. Flowchart of dataset preparation

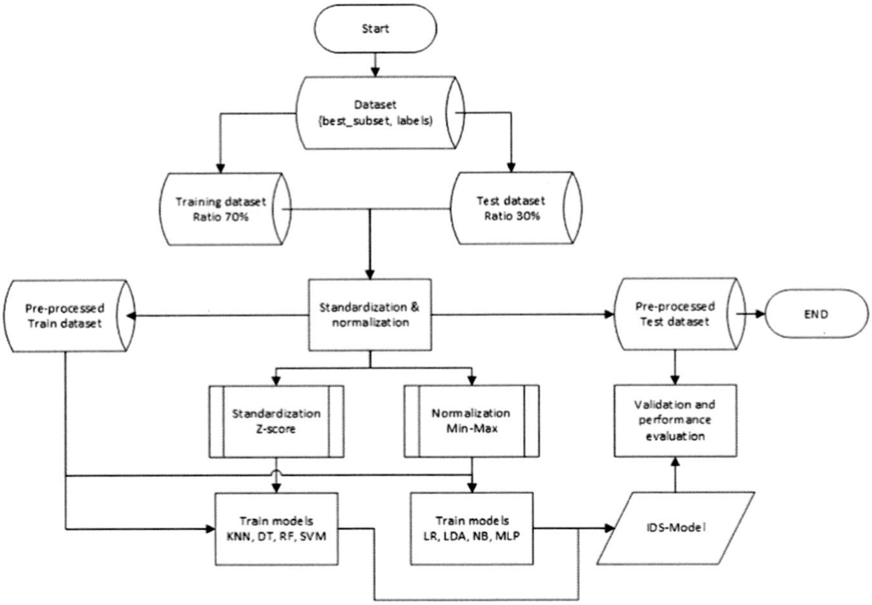

Fig. 2. Flowchart of IDS model building

Table 1. Features categorization

Class	Subclass	CSE-CIC-IDS2018	BoT-IoT
Flow global information	On time and rate features	Fl_dur, fl_iat_avg, fl_iat_std, fl_iat_max, fl_iat_min, ,atv_avg, atv_std, atv_max, atv_min, idl_avg, idl_std, idl_max, idl_min, fl_byt_s, fl_pkt_s, down_up_ratio	Rate, Srate, Drate, AR_P_Proto_P_SrcIP, AR_P_Proto_P_DstIP, AR_P_Proto_P_Sport,AR_P_Proto_P_Dport
	Basic information	subfl_fw_pk, subfl_fw_byt, subfl_bw_pkt, subfl_bw_byt, fw_win_byt, bw_win_byt, Fw_act_pkt, fw_seg_min, fw_seg_avg, bw_seg_avg,	Pkts, Bytes, State, state_number, Ltime, Seq, Dur, Mean, Stddev, Sum, Min, Max, Sbytes, Dbytes, TnBPSrcIP, TnBPDstIP, N_IN_Conn_P_SrcIP, N_IN_Conn_P_DstIP
On packets statistics features	Packets count & size information	fw_pkt_l_min, fw_pkt_l_avg, fw_pkt_l_std, fw_pkt_l_max, Bw_pkt_l_max, bw_pkt_l_min, Bw_pkt_l_avg, Bw_pkt_l_std, pkt_size_avg, tot_l_fw_pkt, tot_fw_pk, tot_bw_pk	Spkts, Dpkts, TnP_PSrcIP, TnP_PDstIP, TnP_PerProto, TnP_Per_Dport, Pkts_P_State_P_Protocol_P_DestIP, Pkts_P_State_P_Protocol_P_SrcIP
	Packets content information	fw_psh_flag, bw_psh_flag, fw_urg_flag, bw_urg_flag, fw_hdr_len, bw_hdr_le, fin_cnt, syn_cnt, rst_cnt, pst_cnt, ack_cnt, urg_cnt, cwe_cnt, ece_cnt	
Global features		fw_byt_blk_avg, fw_pkt_blk_avg, fw_blk_rate_avg, bw_byt_blk_avg, bw_pkt_blk_avg, bw_blk_rate_avg, protocol	Proto, proto_number, pkSeqID, Stime,
Hosts information		Dst_port, Sport	Saddr, Sport, Daddr, Dport

3 Our Method

In this section we present our approach to build an IDS model based on machine learning for botnet detection in classic networks and in the internet of things context.

The proposed methodology to build IDS based on machine learning approaches is summarized in Figs. 1 and 2, the first presents main steps to prepare datasets, when the second demonstrate the pre-processing, models training and validation steps. Our approach can be depicted in the following steps:

1. Data cleaning: The first step to prepare dataset for machine learning is to clean it from outlets, infinite and NaN (not a number) values.
2. features categorization: In this work we will focus on flow based datasets and will propose to categorize their features into 4 main classes:
   - Flow global information features: This category contains features giving global information on each flow, this can be subdivided into 2 other sub-categories:
     - On time and rate based features like flow duration, download and upload rate ratio and so on.
     - Basic information features as number of sub-flows, size of data sent in each window.
   - On packet statistics based features: This category gives information on packets contained in each flow and can be decomposed into 3 sub-classes:
     - Packets number and size information, this contains information like number of packets in the flow, average packet size or standard deviation of packet size in the same flow.
     - On packet content statistics like number of certain flag contained in packets of the flow.
   - Global features like used protocols and any information with values defined in limited interval and common to all flows.
   - Hosts information based features: like IP Address, ports and any information given about source or destination machines.

   In the Table 1 we apply this categorization to the two chosen datasets. Namely CSE-CIC-IDS2018 and BoT-IoT dataset.
3. subsets analysis: This step can be depicted into two phases:
   - Information gain study: In order to determinate each subset of features is the most relevant and brings the most important part of the information we proceed to an ANOVA (analysis of variance) test then train several models with the resulting 10-best features, the resulting accuracy are then compared to our proposed subsets of features. ANOVA is a filter method measuring the relevance of features by their correlation to the target variable instead of cross-over and back-forward between model training, performance evaluation and subset building steps.
   - Correlation study: When the subset is chosen we perform standard correlation test, we retain Pearson for it's simplicity and the fact it is based on variance-co-variance matrix in order to detect correlated features and reduce dimensions.
4. Data sampling and partitioning:
   As explained in Sect. 4.1 we use a simple computer to train models, it was necessary to use only a fraction of 10% of the overall data in order to train and evaluate the different models, the proportion of normal and malicious traffic was saved during the data sampling.

As shown in Fig. 1 the sample was cut into training and testing datasets with ratio of 70% and 30% respectively which is standard in machine learning approach.

5. Standardization and normalization: In order to train the models of the next step, we first standardize and normalize data:
   - Standardization:

$$z - score = \frac{x - \mu}{\sigma}$$

   Where *Mu* is population mean and *Sigma* standard deviation. The z-score expresses values as number of standard deviation between the original observation and the population mean.
   - normalization:

$$Min\_Max = \frac{X - X\mathbf{min}}{X\max - X\min},$$

   this method is used to scale data range between 0 and 1 values which simplify training several models.

6. models training: As shown in Fig. 2, we train several models namely:
   Trained on data normalized with Min-Max function:
   - Logistic regression (LR).
   - Linear Discriminant Analysis (LDA).
   - Naive Bayes (NB).
   - Multi-layer Perceptron (MLP): We trained MLP classifiers with several configuration which are combination of:
     - Activation functions: Logistic, hyperbolic tan, rectified linear unit.
     - Solvers: Quasi-Newton methods optimizer (lbfgs), stochastic gradient descent (sgd) and stochastic gradient-based optimizer (adam).
     - Hidden layers: Two or three with a maximal full number of hidden nodes of 15.
     - Alpha (L2 penality) for regularization term: 1e–5.
     - Adaptative learning rate: This keeps the learning rate constant until failing to decrease training loss between two epochs, then divide learning rate by 5.

   Trained on data standardized with Z-score function:
   - K-Nearest Neighbors (KNN).
   - Decision Trees (DT).
   - Random Forests (RF).
   - Support Vector Machines (SVM): We trained several SVM models varying kernel functions: polynomial, Gaussian, radial basis function (RBF) and sigmoid kernel. The parameter gamma value was fixed to the inverse of number of features when the regularization parameter C was varied between 0.1 to 2.

We don't give a complete explanation of models here, for more details we recommend the paper of Raouf Boutaba et al. [6].

7. Models validation and performance evaluation: At the end of the process, the trained models are tested on the test dataset in order to evaluate in each measure they are able to predict classes and then validate them by performance evaluation metrics described in Sect. 4.2.

We prefer to see our method as a pipeline which can be used to prepare datasets, categorize features and determine best ones then training classification model and evaluate them. This can also be a reference to know what features must be considered for future cybersecurity datasets generation particularly when they are dedicated to botnet detection.

4 Experiments

In this section, we briefly present some of datasets used in the cybersecurity literature, we introduce the metric parameters we use for the evaluation of our models and then give results of experiments and discuss them.

4.1 Environment and Librairies

In order to evaluate performance of our method, we have used sklearn and jupyter notebooks on anaconda environment on a machine with the following features:

- Processor: Intel®Core^{TM} i7-8550U CPU @ 1.80 GHz 1.99 GHz
- RAM: 16 GO DDR4
- OS: Windows 64 bit, x64-based processor

Sklearn is a python library build on NumPy, SciPy and matplot lib, it provides powerful and simple tools for data analysis.

4.2 Evaluation Metrics

To evaluate our models we use the most relevant and commonly used performance indicators namely:

- Accuracy: Expresses the portion of what was correctly predicted.

$$Accuracy = \frac{TP + TN}{TP + FP + FN + TN}$$

- Precision: Evaluates the predicted positive cases related to all what was predicted as positive including negative cases predicted as positive (false positives).

$$Precision = \frac{TP}{TP + FP}$$

- Recall: Contrarily to precision, recall measures the predicted positives related to all real positive cases.

$$Recall = \frac{TP}{TP + FN}$$

- F1-Score: It is a global parameter used to evaluate binary classification models, it is represented by the harmonic mean of recall and precision.

$$f1 - score = 2 * \frac{recall * precision}{recall + precision}$$

- The area under the ROC (Receiver Operating Characteristics) curve: The ROC curve is plotted using true positive rate against false positive rate (Recall), it's a probability curve, then the AUC expresses the degree of separability, in other words it demonstrate the capability of the model to distinguish between predicted classes. When the AUC is near to 1 value then the model is more efficient.
- confusion matrix: the confusion matrix allows summarizing of true positives, false negatives, false positives and true negatives as shown in Table 2.

Table 2. Confusion matrix

Predicted class		1	0
Actual class	1	True Positive	False negative
	0	False positive	True negative

4.3 Results and Discussion

In this subsection we summarize most important results we obtained on the two datasets.

4.3.1 CSE-CIC-IDS Data

The Table 3 presents an overview from accuracy point of view for all trained models on all subsets, an ANOVA based pipeline was implemented too in order to make comparison, it was use-full as a filter method to define the 10 best features. As shown in this table the subcategory of packets count and size globally gives the best results for botnet detection. This is without surprise for the following reasons:

- The number of packets in a botnet flow is less then in a ordinary communication, because it is only in charge of transferring commands and wakes up when an attack is started.
- The average packet size is smaller then standard packets because of the low information volume in the payload part.
- The botnet is not influencing other features like communication rate, flags or anything else.

Table 3. Accuracy based comparison of subsets for evaluating classification models

Model	Flow information		Packets statistics		ANOVA Pipeline
	Time & rate	Basic	Size	Content	
LR	0.735	0.732	0.948	0.739	0.872
LDA	0.735	0.657	0.810	0.743	0.867
NB	0.456	0.791	0.780	0.420	0.783
MLP	0.735	0.929	0.956	0.866	0.876
KNN	0.995	0.987	0.998	0.870	0.884
DT	0.995	0.989	0.999	0.875	0.883
RF	0.980	0.862	0.860	0.862	0.871
SVM	0.735	0.931	0.964	0.865	0.877

For these reasons we retain this subset for further experiments and evaluation.

We first start by performing a Pearson test in order to reveal if there is linear correlation between concerned features, the meaningful results are presented in Table 4 where we can easily establish that standard and maximum packet length are correlated to the mean packet size, this will allow us to reduce the number of features in the selected subset because of the fact that information given by these two features is expressed by the mean value. The pair-plot graph shown in Fig. 3 proves visually the validity of the test.

Table 4. Pearson test matrix for the selected subset

	Pkt Len Min	Pkt Len Max	Pkt Len Mean	Pkt Len Std
Pkt Len Min	1	–0.95	0.11	–0.09
Pkt Len Max	0.05	1	0.84	0.93
Pkt Len Mean	0.11	0.84	1	0.85
Pkt Len Std	0.09	0.93	0.85	1

The metrics given by Table 5 are the mean values of 10 folds training phase, it is dedicated to evaluate performance of botnet detection with different binary classification approaches with the packets number and size subset. The best results for MLP classifier are provided by using adam solver with rectified linear unit activation function and 3 hidden layers composed of 14 nodes.

4.3.2 Bot-IoT Dataset

The Table 6 shows performance evaluation results for botnet detection in the IoT-botnet dataset, we used the same approach and we constat that the selected

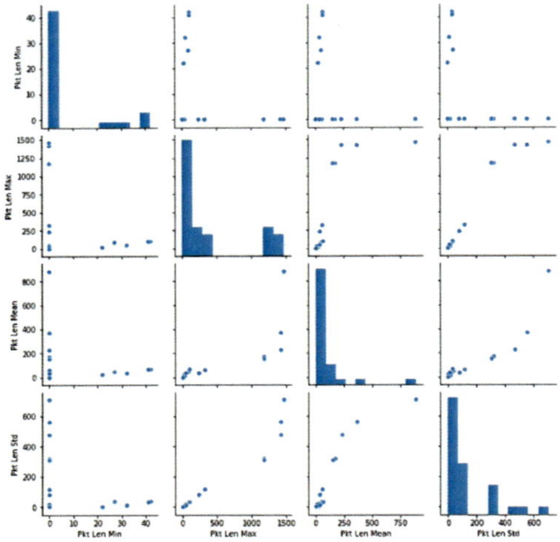

Fig. 3. Pair-plot graph for correlated features

Table 5. Performance evaluation for Botnet detection in CSE-CIC-IDS2018 dataset

		KNN	DT	RFC	SVM	LR	LDA	GNB	MLP
Accuracy		0.987	0.987	0.872	0.952	0.848	0.710	0.788	0.984
Recall		0.99	0.99	0.87	0.95	0.72	0.72	0.81	0.98
F1-score		0.99	0.9	0.86	0.95	0.72	0.72	0.83	0.98
AUC		0.998	0.998	0.963	0.957	0.836	0.782	0.875	0.985
Precision	0	1.0	1.0	0.85	0.97	0.81	0.81	1.00	0.98
	1	0.96	0.96	1.00	0.90	0.48	0.48	0.59	1.00

Table 6. Performance evaluation for Botnet detection in IoT-botnet dataset

		KNN		DT		RFC		SVM		LR		LDA		GNB		MLP	
Accuracy		0.958		0.976		0.964		0.864		0.873		0.870		0.874		0.940	
Recall		0.96		0.98		0.97		0.86		0.86		0.86		0.86		0.92	
F1-score		0.96		0.98		0.97		0.82		0.82		0.82		0.82		0.91	
AUC		0.951		0.973		0.958		0.846		0.695		0.687		0.642		0.922	
Precision	0	0.92		0.96		0.93		0.96		0.92		0.92		0.82		0.96	
	1	0.96		0.98		0.98		0.86		0.86		0.86		0.86		0.92	
Confusion matrix	0	340	72	376	36	364	48	85	327	79	333	81	331	94	318	242	170
	1	31	1952	14	1969	26	1957	4	1979	7	1976	7	1976	20	1963	9	1974

subset gives acceptable detection rate also in the IoT context. Instead of manually trying several configuration, this time we used a grid search approach to optimize the MLP performance due to the lake of normal data in the dataset its accuracy decreases of about 4%. This result was obtained with hyperbolic tan function and with lbfgs solver.

5 Conclusion

In this paper we presented a methodology to build intrusion detection systems based on machine learning, we proposed a categorization of features and compared the obtained results to ones given by a standard features selection methods, namely analysis of variance (ANOVA). This helped us to determine each subset is more relevant for botnet detection, then we trained several models using classic internet dataset: CSE-CIC-IDS2018 and a dedicated one for the internet of things Bot-IoT-dataset. The performed evaluation performance measures proves that features bringing information about number of packets in a flow and their sizes are the most significant for botnet detection in both contexts.

As perspective, we plan to expand our tests to cover all commonly used cybersecurity datasets, we will also consider to build models based on different neural networks architecture like recurrent neural networks (RNN), Long short-term memory (LSTM) or convolutional neural networks (CNN) instead of using only MLP, we will consider to use genetic algorithms in order to find most convenient configuration for hidden layers.

References

1. Abaid, Z., Sarkar, D., Kaafar, M.A., Jha, S.: The early bird gets the botnet: a markov chain based early warning system for botnet attacks. In: 2016 IEEE 41st Conference on Local Computer Networks (LCN), pp. 61–68. IEEE (2016)
2. Abdul Kadir, A.F., Stakhanova, N., Ghorbani, A.A.: Android botnets: what urls are telling us. In: Qiu, M., Xu, S., Yung, M., Zhang, H. (eds.) Network and System Security. NSS 2015. LNCS, vol. 9408. Springer, Cham (2015). https://doi.org/10.1007/978-3-319-25645-0_6
3. Raza, S., Wallgren, L., Voigt, T.: SVELTE: real-time intrusion detection in the internet of things. Ad Hoc Netw. **11**, 2661–2674 (2013)
4. Almutairi, S., Mahfoudh, S., Alowibdi, J.S.: Peer to peer botnet detection based on network traffic analysis. In:2016 8th IFIP International Conference on New Technologies, Mobility and Security (NTMS), pp. 1–4. IEEE (2016)
5. Bezerra, V.H., Da Costa, V.G.T., Barbon Junior, S., Miani, R.S., Zarpelão, B.B.: IoTDS: a one-class classification approach to detect botnets in internet of things devices. Sensors (Switzerland) **19**(14), 1–26 (2019)
6. Boutaba, R., et al.: A comprehensive survey on machine learning for networking: evolution, applications and research opportunities. J. Internet Serv. Appl. **9**(1), 1–99 (2018). https://doi.org/10.1186/s13174-018-0087-2
7. Dorsemaine, B., Gaulier, J.P., Wary, J.P., Kheir, N., Urien, P.: A new approach to investigate iot threats based on a four layer model. In: 2016 13th International Conference on New Technologies for Distributed Systems (NOTERE), pp. 1–6. IEEE (2016)
8. Ferrag, M.A., Maglaras, L., Moschoyiannis, S., Janicke, H.: Deep learning for cyber security intrusion detection: approaches, datasets, and comparative study. J. Inf. Secur. Appl. **50**, 102419 (2020)
9. Beigi, E.B., Jazi, H.H., Stakhanova, N., Ghorbani, A.A.: Towards effective feature selection in machine learning-based botnet detection approaches. In: IEEE Conference on Communications and Network Security (CNS) (2014)

10. Sharafaldin, I., Lashkari, A.H., Ghorbani, A.A.: Toward generating a new intrusion detection dataset and intrusion traffic characterization. In: Proceedings of the 4th International Conference on Information Systems Security and Privacy - ICISSP (2018)
11. Hodo, E., et al.: Threat analysis of iot networks using artificial neural network intrusion detection system. In: 2016 International Symposium on Networks, Computers and Communications (ISNCC) (2016)
12. Julesa, J.M., Cheng, H., Regedzaic, G.R.: A survey on botnet attacks. Am. Sci. Res. J. Eng. Technol. Sci. (ASRJETS) **77**(1), 76–89 (2021)
13. Koroniotis, N., Moustafa, N., Sitnikova, E.: Forensics and deep learning mechanisms for botnets in internet of things: a survey of challenges and solutions. IEEE Access **7**, 61764–61785 (2019)
14. Koroniotis, N., Moustafa, N., Sitnikova, E., Turnbull, B.: Towards the development of realistic botnet dataset in the Internet of Things for network forensic analytics: Bot-IoT dataset. Futur. Gener. Comput. Syst. **100**, 779–796 (2019)
15. Lin, K.-C., Chen, S.-Y., Hung, J.: Botnet detection using support vector machines with artificial fish swarm algorithm. J. Appl. Math. **2014**, 1–9 (2014). https://doi.org/10.1155/2014/986428
16. Larriva-Novo, X.A., Vega-Barbas, M., Víctor, V.A., Rodrigo, M.S.: Evaluation of cybersecurity data set characteristics for their applicability to neural networks algorithms detecting cybersecurity anomalies. IEEE Access **8**, 9005–9014 (2020)
17. Mahdavifar, S., Kadir, A.F.A., Fatemi, R., Alhadidi, D., Ghorbani, A.A.: Dynamic android malware category classification using semi-supervised deep learning. In: Proceedings - IEEE 18th International Conference on Dependable, Autonomic and Secure Computing, IEEE 18th International Conference on Pervasive Intelligence and Computing, IEEE 6th International Conference on Cloud and Big Data Computing and IEEE 5th Cyber Science and Technology Congress, DASC/PiCom/CBDCom/CyberSciTech 2020, pp. 515–522 (2020)
18. Slay, J., Moustafa, N.: UNSW-NB15: a comprehensive data set for network intrusion detection systems (UNSW-NB15 network data set). In: 2015 Military Communications and Information Systems Conference (MilCIS) (2015)
19. Parmisano, A., Garcia, S., Erquiaga, M.J.: A labeled dataset with malicious and benign iot network traffic. Stratosphere Laboratory, Praha, Czech Republic (2020)
20. Ring, M., Wunderlich, S., Scheuring, D., Landes, D., Hotho, A.: A survey of network-based intrusion detection data sets. Comput. Secur. **86**, 147–167 (2019)
21. García, S., Grill, M., Stiborek, J., Zunino, A.: An empirical comparison of botnet detection methods. Comput. Secur. **45**, 100–123 (2014)
22. Sarker, I.H., Kayes, A.S.M., Badsha, S., Alqahtani, H., Watters, P., Ng, A.: Cybersecurity data science: an overview from machine learning perspective. J. Big Data **7**(1) (2020)
23. Spathoulas, G., Giachoudis, N., Damiris, G.P., Theodoridis, G.: Collaborative blockchain-based detection of distributed denial of service attacks based on internet of things botnets. Futur. Internet **11**(11) (2019)
24. Tavallaee, M., Bagheri, E., Lu, W., Ghorbani, A.A.: A detailed analysis of the kdd cup 99 data set. In: 2009 IEEE Symposium on Computational Intelligence for Security and Defense Applications, pp. 1–6. IEEE (2009)
25. Vinayakumar, R., Alazab, M., Srinivasan, S., Pham, Q.V., Padannayil, S.K., Simran, K.: A visualized botnet detection system based deep learning for the Internet of Things networks of smart cities. IEEE Trans. Ind. Appl. **56**(4), 4436–4456 (2020)
26. Xia, H., Li, L., Cheng, X., Cheng, X., Qiu, T.: Modeling and analysis botnet propagation in social internet of things. IEEE Internet Things J. **7**(8), 7470–7481 (2020)

Detecting Vulnerabilities in Source Code Using Machine Learning

Omar Hany[(✉)] and Mervat Abu-Elkheir

German University in Cairo, Cairo, Egypt
omar.mohamedgaber@student.guc.edu.eg, mervat.abuelkheir@guc.edu.eg

Abstract. In recent years, software vulnerabilities have been the source of countless cyber attacks. Despite the existence of various methods for detecting software vulnerabilities like static analysis tools and dynamic analysis tools, the number of vulnerabilities discovered each year continues to climb. Over the last decade, machine learning models have made a significant progress. As machine learning methods, do not need human experts to define features and can learn vulnerability patterns automatically as it captures patterns that human may not understand. The goal of this paper is to create a machine learning model that could discover vulnerabilities in the function scope (i.e. method or procedure in any programming). To accomplish this aim we propose a novel feature extraction technique based on clustering the vocabulary of the function text using Kmeans. Typically, the vulnerability classification problem is an imbalanced one, as most functions are naturally not vulnerable. We use a dataset and compare the effect of four different class imbalance handling techniques (Class weight, Random Undersampling, Random Oversampling and Synthetic Minority Oversampling Technique (SMOTE)). Results show that using the Class Weight modification technique is the best in both metrics we use (Recall and F1 score) (76% and 17%). The results show that our method has very comparable results relative to other methods in addition to faster training time due to using shallow machine learning model.

1 Introduction

Computers have become a vital part of today's generation's lives; from instant messaging and email to banking, travel, education, and shopping, computers have impacted every element of existence. Protecting sensitive information has become a need as people's use of computers has increased. The chances of cyber exploitation and cyber crimes are increasing as the number of data networks, digital apps, and internet and mobile users grows. These cyber attacks has a great impact on economy as it is predicted that cyber attacks in 2021 only will cause a loss of $6 trillion and it is expected to be $10.5 trillion by 2025 [1]. The main cause of these cyber crimes and information leakages are software vulnerabilities. Software vulnerabilities are flaws in programs that make software programs do not work in the way they are designed to. Despite the fact that several ways to

© The Author(s), under exclusive license to Springer Nature Switzerland AG 2022
H. Ragab Hassen and H. Batatia (Eds.): ACS 2021, LNNS 378, pp. 35–41, 2022.
https://doi.org/10.1007/978-3-030-95918-0_4

detecting software vulnerabilities have been presented like static analysis tools which examine security flaws in code without running it and dynamic analysis tools which find vulnerabilities by investigating the properties of a program using information gathered at run-time, the number of vulnerabilities revealed each year continues to rise as the overall number of new vulnerabilities in 2019 (20,362) increased by 17.6% compared to 2018 (17,308) and by 44.5% compared to 2017 (14,086) [2]. The problem software vulnerability and cyber attacks will continue to exist for a long time. Therefore, more efficient and accurate solutions are needed.

Over the last decade, both artificial intelligence (AI) as a broad field and machine learning as an AI tool have made significant progress. Many data-driven AI systems that ingest vast amounts of data and learn an accurate model of the underlying distribution. They also have the advantage of not requiring explicit extraction of the features that are most necessary to define the model. Machine learning frequently uncovers hidden aspects that a human handcrafted model may not consider. The advancement of machine learning technology provides new approaches for detecting vulnerabilities. As machine learning methods, do not need human experts to define features and can learn vulnerability patterns automatically and also it can capture patterns that human may not understand or capture.

Previous works based on machine learning approaches to detect vulnerabilities in source code function is categorized into two main types. The first type is Graph-based representation [3,4] where the different graphs are extracted from the source code functions like Abstract Syntax Trees (AST), Control Flow Graph (CFG) and Data Flow Graph (DFG) and then these graphs are flattened and fed to a deep learning model. The second type is Sequence-based representations [5] in which sequences are extracted from the source code text tokens and then fed to deep learning classification models. Both these approach have the two shortcomings: (1) Large model size (2) Long training and inference times.

In this work, we propose a novel feature extraction technique from source code text in which we cluster the vocabulary of source code and record the frequency of the tokens that belongs to each cluster in a given function. The objective of using clustering is to generate semantic rich features where each token is represented by the cluster it belongs to. We hypothesized that tokens that belong to same cluster will have a similar semantics in the function (e.g. variables will typically be clustered together). Having each function represented by the frequency of tokens belonging to each cluster an effective shallow machine learning approach, namely, Random Forest is used to classify the function as either vulnerable or not. Due to the imbalanced nature of vulnerable functions relative to non vulnerable function which is reflected in the dataset we use where the number of vulnerable functions are far less the non vulnerable functions we apply four types of imbalance handling techniques and compare their results.

2 Related Work

Current approaches to detect vulnerabilities in source code text can be categorized into two categories: (1) Graph-based feature representation (2) Sequence-based feature representation. As for the Graph-based feature representation, Zhou et al., 2019 [4] believed that source code is actually more structural and logical than natural languages. They extracted Abstract Syntax Tree (AST), Data Flow Graph (DFG) and Control Flow Graph (CFG) from the source functions using open source platform. The concatenated the extracted graphs after flattening and fed it to a CNN deep learning model. Also, S. Liu et al., 2020 [3] extracted Abstract Syntax Tree (AST) from source code functions using an open source program and fed it to a Long Short Term Memory (LSTM) deep learning model. On the other for the Sequence-based representation approach, Russell et al., 2018 [5] extracted features from the source code they created a custom C/C++ lexer designed to reduce C/C++ code to representations using a total vocabulary size of only 156 tokens to build the classifier models. For each function they created a (500×13) vector that represents the token inside the function. They used CNN deep learning model as a classification model.

Having discussed the related work have not been able to encode the semantic into a data structure of a manageable size. Hence, such models are typically large and slow to train with. In this work we propose a semantic rich approach utilizing the strengths of word embeddings, more specifically Word2Vec. The proposed approach builds on Word2Vec to generate a novel feature extraction method by clustering the vocabulary of each function and use the frequency of cluster instance occurrence as a proxy for semantics (i.e. tokens to belongs to same cluster are likely to be semantically similar).

3 Background

3.1 Class Imbalance

Predicting the label of a record is the goal of machine learning classification models. An imbalanced classification problem is one in which the distribution of examples across recognised classes is uneven or biased. The distribution can range from a little skew to a severe imbalance, with one example in the minority class for hundreds, thousands, or millions in the majority class.

Imbalanced classifications present a difficulty for classification models, as most machine learning algorithms were built with the assumption of an equal amount of samples for each class. As a result, models with poor prediction accuracy, particularly for the minority class, emerge. This is a problem because the minority class is typically more significant than the majority class, and hence the problem is more susceptible to categorization errors in the minority class than in the majority class.

It's possible that the imbalance is a characteristic of the problem domain as in our problem of detecting vulnerabilities which has the nature of being rare compared to non-vulnerable codes. For example, one class's natural occurrence or presence may pre-dominate over others. This could be due to the fact that the procedure for generating observations in one class is more costly in terms of time, money, computing, or other resources. As a result, just collecting more samples from the domain to increase the class distribution is often impractical or impossible. There are ways to limit the effect of the class imbalance problem such as (1) Class weight modification (2) Random undersampling (3) Random oversampling (4) SMOTE.

Class Weight Modification: Class weights in binary classification could be modified simply by calculating the weight of the minor and major classes and then inverting it so that the minor class has a considerably bigger error than the majority class when multiplied by the class loss. As the minor class is the one we are more interested in.

Random Undersampling: Removal of random samples from the majority class. This is one of the first strategies for correcting dataset imbalances; nonetheless, it may raise the classifier's variance and is extremely likely to remove relevant or crucial samples.

Random Oversampling: Adding random extra copies of some of the minority classes to the training data. This is one of the first approaches proposed, and it has been demonstrated to be reliable. Rather of replicating every sample in the minority class, some of them might be picked at random and replaced.

SMOTE: The minority data is duplicated from the minority data population in traditional oversampling process. While it increases the amount of data available, it does not provide the machine learning model with any new information or variation. SMOTE is a type of oversampling in which synthesized new instances of the minority classes are added to the training data. SMOTE works by utilizing a k-nearest neighbor algorithm to create synthetic data.

3.2 Random Forest

RF is an ensemble learning method for classification and regression. Breiman (2001) [6] developed the method, which combines Breiman's bagging sampling methodology [7] with Ho (1998) and Amit and Geman (1997)'s random selection of features to create a collection of decision trees with controlled variation. Each decision tree in the ensemble is built via bagging, which involves replacing a sample with data from the training set. According to statistics, around 64% of the instances in the sample will appear at least once. These instances are called inbag instances. While the remaining instances are called out-of-bag instances make up the remaining 36%. To derive the class label of an unlabeled instance, each tree in the ensemble works as a base classifier. This is accomplished through majority voting, in which each classifier casts one vote for its anticipated class label, with the most votes being used to categorise the instance.

3.3 Word2Vec

Word2Vec model is a collection of linked models for generating word embeddings. Word2Vec has shallow two-layer neural networks that have been trained to recreate word linguistic contexts. Word2vec takes a big text as input and outputs a vector space that reflects the semantic behind the text. Word vectors are positioned in vector space in such a way that words with similar contexts in the text are close to one another and have similar values. For example, in our case variables will have close values and so do function names.

4 Proposed Methodology

The proposed methodology of this paper is mainly a semantic rich approach utilizing the strengths of word embeddings, more specifically Word2Vec. The proposed approach builds on Word2Vec to generate a novel feature extraction method by clustering the vocabulary of each function by Kmeans and use the frequency of cluster instance occurrence as a proxy for semantics (i.e. tokens to belongs to same cluster are likely to be semantically similar).

Having each function represented by the frequency of tokens belonging to each cluster an effective shallow machine learning approach, namely, Random Forest is used to classify the function as either vulnerable or not. Due to the imbalanced nature of vulnerable functions relative to non vulnerable function which is reflected in the dataset we use where the number of vulnerable functions are far less the non vulnerable functions we apply four types of imbalance handling techniques (Class Weight modification, Random Undersampling, Random Oversampling and SMOTE [8]) and compare their results.

-

5 Experimental Study

5.1 Dataset

D2A [9] is a dataset created by IBM to train models for vulnerability identification. The dataset is labeled by a tool called D2A which labels records more accurate than any static analysis tool and decrease the amount of false positives as it runs a static analysis tool on two versions of the same program. If a program is triggered to be vulnerable by the static tool the program is labelled vulnerable only if the vulnerable part disappeared from the newer version of the program. The dataset contains more than 10 types of famous different vulnerabilities that C/C++ programs have the most, like the Buffer overflow and Null pointer exception vulnerabilities. The dataset contains real-world C/C++ functions extracted from six large programs (Openssl, FFmpeg, httpd, NGINX, libtiff and libav). Due to computational limitations we will use only NGINX program. NGINX has a total number of 18,366 functions. Class distribution is very imbalanced as shown in where vulnerable functions are only 421 functions (2.2%) while non vulnerable functions are 17,945 functions (97.8%).

5.2 Training

As our approach is a token based clustering approach to the vocabulary of the source code functions, we perform tokenization over the whole vocabulary of all functions of the given dataset. then, we perform the feature extraction process to score each token of the source code text. we use Gensim Word2Vec to generate a vector representation of the tokens. After that, we use distributed the vectorized tokens into k clusters based on the clustering algorithm K-Means where the number of clusters is 50. For each function we end by a vector of size 50 and for each index it donates the number of tokens that belong to the cluster of this index in a given function. Due to the imbalanced nature of vulnerable functions relative to non vulnerable function which is reflected in the dataset we use where the number of vulnerable functions are far less the non vulnerable functions we apply four types of imbalance handling techniques (Class Weight modification, Random Undersampling, Random Oversampling and SMOTE [8]). Then, we use a shallow machine learning model, namely, Random Forest on the generated vectors. The number of trees for the Random Forest classifier is tuned to be 1000. As for the Class weight technique the class weight is calculated using scikit learn library and given as a parameter to the Random Forest model. Then, we obtain results and compare them.

5.3 Results

We performed experiments on a device running on 32 GB of RAM, Intel(R) Xeon(R) CPU E5-1603 v3 processor and Nvidia GTX 780ti gpu. We used Random Forest classifier using scikit learn library for the output vectors. The experiment was repeated the imbalanced dataset and after addressing the issue using four different approaches: (1) Class weight modification (2) Random Undersampling (3) Random Oversampling (4) SMOTE. We evaluated the model using Recall and F1 score metrics as they best reflect the performance of imbalanced datasets. Results obtained are shown in Table 1.

Table 1. A comparison among the between our method and Russel et al. using different class imbalance treatment approaches

Technique	Imbalanced		Class weight		Random Undersample		Random Oversample		SMOTE	
	Recall	F1	Recall	F1	Recall	F1	Recall	F1	Recall	F1
Novel Feature extraction + RF	0	0	**76%**	**17%**	75%	10%	**76%**	14%	74%	15%
Russel et al. CNN	0	0	**78%**	**24%**	74%	11%	60%	22%	40%	20%

We compared our feature extraction technique with the Random Forest classifier with applying the CNN architecture of Russel et al. [9]. The results show that when the class imbalance problem has not been addressed all used metrics has a zero value as the model tends to predict only the major class (non Vulnerable) because the model optimizes for accuracy. Our method performed nearly as good as the Russel et al. architecture in the Class Weight modification method. Specifically, the recall metric is (76% and 78%) for our method and Russel et al. method respectively. While in the F1 score value our method scored 17% relative to the 24% in Russel et al. From the results it is clear that the class weight modification method is the best method to addressing the problem of the class imbalance. It is also clear that our method is effective as the size of the generated vector for our feature extraction is far less than Russel et al. ((50×1) vs (500×13)) and that is reflected in the faster training time.

6 Conclusion and Future Work

The aim of this paper is to propose a new method with fast training time achieving a comparable predictive accuracy. To accomplish this aim we proposed a novel feature extraction technique using Kmeans and trained a Random Forest Classifier with a C/C++ source code functions labled dataset with famous vulnerabilities types. As the vulnerability classification problem is an imbalanced one which means most functions are naturally not vulnerable, we used four class imbalance handling techniques (class Weight, Random Undersampling, Random Oversampling and SMOTE). We compared the results of the four techniques. The results show that using the Class Weight modification technique is the best in both metrics we use (Recall and F1 score) (76% and 17%). The results also show that our method has very comparable results relative to other method. Typically the resulted model is faster in trainig due to the shallow machine learning methodology adopted.

Future directions of research include experimenting the variants of the number of clusters and using other clustering methods than Kmeans.

References

1. Sausalito, C.: Cybercrime to cost the world $10.5 trillion annually by 2025 (2020). https://cybersecurityventures.com/hackerpocalypse-cybercrime-report-2016
2. Bekerman, D., Yerushalmi, S.: The state of vulnerabilities in 2019 (2020). https://www.imperva.com/blog/the-state-of-vulnerabilities-in-2019
3. Liu, S., Lin, G., Han, Q.-L., Wen, S., Zhang, J., Xiang, Y.: Deepbalance: deep-learning and fuzzy oversampling for vulnerability detection. IEEE Trans. Fuzzy Syst. **28**(7), 1329–1343 (2020)
4. Zhou, Y., Liu, S., Siow, J., Du, X., Liu, Y.: Devign: effective vulnerability identification by learning comprehensive program semantics via graph neural networks, September 2019
5. Russell, R., et al.: Automated vulnerability detection in source code using deep representation learning, pp. 757–762 (December 2018)
6. Breiman, L.: Random forests. Mach. Learn. **45**(1), 5–32 (2001)
7. Breiman, L.: Bagging predictors. Mach. Learn. **24**(2), 123–140 (1996)
8. Nitesh, V.C., Kevin, W.B., Lawrence, O.H., Kegelmeyer, W.P.: Smote: synthetic minority over-sampling technique. J. Artif. Int. Res. **16**(1), 321–357 (2002)
9. Zheng, Y., et al.: D2A: A Dataset Built for AI-Based Vulnerability Detection Methods Using Differential Analysis. arXiv e-prints, page arXiv:2102.07995 (February 2021)

Android Malware Detection Using Long Short Term Memory Recurrent Neural Networks

Lilia Georgieva(✉) and Basile Lamarque

Heriot Watt University, Edinburgh EH14 4AS, UK
{L.Georgieva,bl2004}@hw.ac.uk

Abstract. In this paper, we study the security attacks on Android using Long Short Term Memory (LSTM) Recurrent Neural Networks. As one of the most popular operating systems, Android is a prime target for security attacks. Only in 2019, 10.5 million malware was detected. Recursive neural networks are in essence machine models made up of a list of cells. Their particularity is that part of the output of the previous cell is used as input for the next one. LSTM have shown good results in several areas, for example, text generation, translation, trajectory prediction. Among the recursive neural network models, LSTM is one of the most efficient approaches to sequence classification as it is able to make relations between very distant elements in a sequence. This research explored the application of LSTM for Android malware detection using source code decompiled from the Android Application Package (APK). The approach we have tried is to first extract the instructions from the source code while respecting their execution order as much as possible. We then explored several ways to filter and encode these instructions. For all feature sets we created, we obtained an accuracy greater than 70% of accuracy and for some feature sets the accuracy reached 83% showing that it is possible to successfully detect malware using source code and LSTM.

1 Introduction

In 2004 six antivirus companies received a piece of code written in C++ [1]. However, only Roman Kuzmenko working at Kaspersky recognized the nature of this code as malware. This was the first mobile phone virus in history. Even if this virus did not do any significant action on the phone (it only displayed Caribe when the phone turns on) it served as a proof of concept that malware can be developed on mobile. Shortly afterwards, a new malware was detected, this time this was a Trojan present in a version of a mobile game. Every time the game started, the phone sent a premium (paid) text message to a phone

We gratefully acknowledge input from Hani Ragab which has been invaluable in this submission.

© The Author(s), under exclusive license to Springer Nature Switzerland AG 2022
H. Ragab Hassen and H. Batatia (Eds.): ACS 2021, LNNS 378, pp. 42–52, 2022.
https://doi.org/10.1007/978-3-030-95918-0_5

number. This became the first mobile virus to steal money from their victims. This is the official start of mobile malware.

The increase in popularity of modern smartphones and their App market combined with the explosion of the Internet, leads to more attacks on mobile devices which are relevant to increasingly higher number of users. Android which is the most popular operating system ahead of Windows with 40% of the market share [2]. A malware attack could, for example, use the identifiers of your email address to do phishing, transfer money through your banking application, install a crypto-currency miner or simply send text messages to an over-taxed number. The development of alternative or unofficial application markets, which are less secure makes the distribution more numerous and significantly easier. In addition, with smartphones integrated into Internet of Things, the number of malware in different platforms is increasing, with 10.5 million malware detected during the year 2019 on Android [3]. Furthermore as with computer malware, signature malware scans used by free antivirus software are often ineffective [4] as their effectiveness generally depends on the database they rely on. To ensure security of the devices, several approaches have been used, for example, static or dynamic analysis. However, these laborious techniques require a lot of time. With over 100,000 apps being developed over the course of only a month, checking the security of every app is not feasible [5]. This is why we explore an alternative approach by creating a deep learning model and show that with a reasonable degree of accuracy we can cover huge volume of submitted applications.

2 Related Work

Android was designed and envisioned as a universal operating system. It is an open source operating system which takes care of managing the basic functionality of the device on which it runs (for example telephone, address book, calendar). It is highly customizable; each manufacturer can add the features they need. This has allowed Android to be installed on phones and tablets, but also on connected devices like televisions or cars, which makes it the most popular operating system in the world. This popularity makes it a prime target for hackers. Research into the detection of android malware is therefore a subject widely explored because of the needs of the industry. The first thing to know about the APK format is that it is actually a zip archive. It is therefore very easy to extract the content of the application by changing its extension and using any compression software (winrar for example) to extract its content.

An APK contains [6]:

- **META-INF**/: folder which contains the manifest information and other metadata about the Java package carried by the jar file.
- **lib**/: libraries that the app uses.
- /: Resources that were not compiled into resources.arsc such as images.
- **assets**/: All the resources provided by the developer that the app may need.
- **AndroidManifest.xml**: Mandatory for every app, this file contains all the meta data that Android needs to know to run your app such as the package name, the permission needed, or the hardware needed.

- **classes.dex**: Compiled Java code.
- **resources.arsc**: Compiled resources, such as Strings.

Typical methods for malware analysis on Windows are divided into 2 types: static analysis and dynamic analysis.

Static analysis is the analysis of the different files that make up a software (binary, resources etc.) to determine the utility of it. The software is never launched during a static analysis. The static analysis of an Android application is therefore done by analyzing the content of a APK. AndroidManifest.xml is a very important file for the analysis, it contains the name, the version but also all the authorization and the hardware required by the application. The simple presence of a certain authorization can tell us that the application is malicious, indeed, for example, a calculator app should not require to have access to your contacts to work as well it should not need that the phone is equipped with a GPS. In addition, it is common for malware to be much more likely to require more authorization than safe applications with an average of 12.99 authorizations needed for malware against 4.5 for safe applications [7].

We explore recent advances of machine learning using as a base the research of [8], which extracts the requested authorization, the hardware required but also the Activity names. These features are initially associated with several machine learning algorithms (C4.5 Decision trees, JRIP or AdaBoost) but also clusterization algorithms (SimpleKMeans, FarthestFirst, EM) and show good results (accuracy > 75%). These good results show the importance of these features and will be reused by other studies which will complete them to other features [9–11]. If we consider the APK file classes.dex, which contains all the compiled code which will be used by Android to execute our application we note that this code can be decompiled in several forms using different tools (APKtool, dex2smali, dex2jar). The decompiled code is full of information that could be useful during an analysis, however its complex nature limits the information that can be easily extracted from it. So research is mainly interested in specific parts of the decompiled code. In [12] it is the frequency of API calls that are extracted from the decompiled code. API calls are functions provided by Android to perform certain actions so that developers do not have to integrate them. Other pieces of code that may be of interest are the Intent, they are used to communicate with another application, for example to set an alarm [13] but it is also possible to receive information using Intent. We can, make our application perform certain actions when the phone is in a certain state, for example just after booting. We note that to perform these actions, they must be specified in the AndroidManifest.xml. In [11] adding intents to the features pools (permissions and API calls) increases the precision of the different tested models. However, such static analysis has certain limitations: information (such as foe example source code or authorization) could be present without ever actually being used, or the application could download code to run it during the execution and this code will not be covered by a static analysis. As an alternative, dynamic analysis, which looks at how the software is running as well as how it influences the machine once launched can be used. Dynamic analysis is usually performed after a static analysis when all possibilities have been completed or when you want to learn more about how the malware works. Unlike a static analysis, a dynamic analysis allows you to observe the malware in operation. This also makes it possible to invalidate or validate the theories made during the static analysis, for example a domain name could be present without being used by the application just like a function could be present in the code without actually being used. Although dynamic analysis shows good results in detecting malware, it has a significant limitation: it is possible that some function of the application are only triggered when a list of events is performed. Although there are tools to generate a random event list and consequently perform dynamic analysis on the application [14] this limitation makes automated dynamic analysis much more time-consuming.

3 Implementation

In this section we present our dataset and the feature extraction and data embedding which are the backbone of our approach.

3.1 Data Set

The database is made up of 2 sources: virusshare.com [15] for malware and androzoo.uni.lu [16] for benign applications. From these two sources we have downloaded a large number of applications before selecting certain applications to create our dataset. The training set is composed of 200 applications, 100 malware and 100 benign, the test set is composed of 50 applications, 25 malware and 25 benign. When selecting the applications we tried to make our dataset as complete as possible, each malware has been analyzed by virustotal.com in order to know their objective. We then created a dataset including as many types of malware as possible. You can find the proportion of each type of malware in Table 1.

Table 1. Malware type representation in training set and test set

Malware type	Number in training set	Test set
Trojan	53	13
Ransomware	23	7
Password Trojan	19	14
Other	5	1

3.2 Feature Extraction

For the features extraction, we will closely follow the idea that a software is only a series of instructions. We will find the starting points of the program then we will follow the execution path and we are going to abstract each line of the code. This method have two main advantages: First, because we are following the code execution path, we are going extract only methods and lines that are actually executed which prevents our model to look in functions that are never called. Secondly, by creating the feature this way, we precisely preserve the "sequence" propriety of the source code because each line that we will be extracting will be after the line which precedes it and before the line which will follow in the execution process. However the code of a mobile application is rarely linear, indeed the presence of a user interface (UI) being almost mandatory for a mobile application, means that certain functions are only called if certain events occur (for example click on a button). The functions that are only called if an event occurs are called callback and the functions that setup this callback are called listener. These methods are therefore quite complicated to place in the general code execution since their execution depends on the user but also on the state of the application. The approach that has been chosen is to include them after the end of the function that sets the listener. Therefore, we considered that as soon as a listener is created, it is called. We chose this approach to reflect that the hacker has an interest in having the malicious code run independently of user input. Note that unlike a classic code, an APK can have several entry points. Indeed thanks to the service and the receivers, certain part of the code can be called only when the phone is in a certain state (just after switching the phone on for example) and even the pages of the application can behave differently if they are launched for the first time or if one returns to the application without having closed it.

3.3 Data Embedding

The objective of data embedding is to transform each instruction which is in the form of a string (for example: move v1, v2) into a vector of fixed size which can be passed to LSMT model while retaining as much information as possible. To do this, we decided to divide all the instructions into 3 pieces, opcode, registers and variables.

The APK manager decompiles the .dex file into .smali and binary .xml to human readable xml. All smali code instructions always start with an opcode, so this is the first element, which we will normalize. Because an opcode is a string that describes what this line is going to do, we will create a vector with size equal to the number of opcode that exits and simply assign each of the cells of that vector to an opcode. For example the first cell will be for the move opcode so if the opcode present on a line is move, the first cell will be at 1 and all the others at 0. This method is called the one-hot embedding.

The registers are the second part that we will extract from the line. Most opcodes use registers to function and they are always formatted in same way.

To encode this part we are also going to do a one-hot embedding. We are going to create a vector of arbitrary size, and if a register is used in this line then the index of this register value will be 1. If ever the value of the register exceeds the size of the array it will simply be ignored. Note that unlike for the opcode, because an instruction could need multiple register, several indexes of the array can be at 1 simultaneously. The size of the array being set arbitrarily, this makes it a hyper parameter.

The variable part represents everything that is not a register or an opcode. It can therefore be a string, an int, a float, or even a function name. This diversity makes this part very hard to embed. Indeed it's very hard to embed string and int. In addition to the diversity of type, the names as well as the number of functions differ from one application to another. The approach that we have made is to consider only the values which part of the Android constants and the function names which are API calls. Indeed this information does not vary from one application to another and can be retrieved from the Android documentation. We have therefore created a python script that will automatically fetch all Android documentations page to retrieve all the constant but also the public methods. Android being extremely complex, we have retrieved 44213 items from the documentation. In the same way as for the opcode, we therefore do a hot-one embedding on an array of 44213.

4 Feature Set Creation

We wanted to keep as much information as possible, which mean keeping all the instructions but also all the information that we had by lines. However, we reach a quantity of data which is too large. If we are able to extract 34,018 instructions from an apk sample, this 34,018 have to be multiplied by 44,823 which is the size of our feature (opcode array + 200 registers array + Variables with Android constants and API calls array) for each row which gives a final array of 1,524,788,814 items. Such a number of elements will not be effectively computable by any models and will even overload the RAM if we try to load several samples at the same time. We first try to reduce the instruction size by reducing is the sizes of the array relative to the variable. Indeed this one comprising more 40,000 entries, even if Title Suppressed Due to Excessive Length 7 there are only 100 instructions, that already represents 4,000,000 tables entries. The first method we did was to change the encoding from hot-one embedding on APIs and constant android to hot-one embedding on the variable type. So we create an array of size 8 representing all types: "cond", "function", "string", "goto", "object", "array", "switch", "try". All elements of the array are 0 except the corresponding type. However, it is obvious that with this method we lose a lot of information. The second method that we tried to use in order to reduce the size of the array of variables was to go through the code and note the percent and the proportion of each of the API call and constant. So for our dataset we had a features set representing the proportion of use of each of the elements present in the Android array of variables. We then applied several feature selection algorithms to reduce the size of the array (see result Table 2). With the result of these algorithms we built a new array for our variables.

Table 2. Result for the different feature selection algorithms (number of features at the beginning 44211)

Feature selection	Output array size	Hyper parameter
One-hot embedding on type	6	NAN
Variance	1192	Threshold = 0.2
Variance	1001	Threshold = 0.3
Variance	946	Threshold = 0.4
Select model from	1783	Logistic regression
Select model from	1783	SGD classifier
Select percentile	4469	Percentile = 10

To reduce the instruction number, we keep only the instructions that call an API function and constant. In this way our data will represent their presence but also their proportion in addition to the order in which they are called and this will also greatly reduce the number of instructions that we will extract.

Another way to reduce the number of instructions is to limit the depth to which we explore functions. We will therefore define an N as our maximum depth, and the starting point as the depth 0. A method called by the starting point will have a depth of 1 and a method called by this method will have a depth of 2 since it will be 2 call from the starting point. Note that if the starting point calls several functions, all these functions will have a depth of 1 since they will only be one call from the starting point. When a function reaches the depth N, it is considered that this function is empty and therefore it's not explored or extracted either. By doing this, we reduce the percentage of code covered, however the code that is kept is a code that is close to the starting points and which therefore has a better chance of having been written by the publisher (see Table 3).

We next chose to create four features sets from the different selection methods.

- Reference: This set of features will be created with a one-hot embedding for the opcode as well as for the register. However, the size of the array assigned to the register will be 10 (5 for registers and 5 for parameters). For the variable, we are going to use a one-hot embedding on the variable filtered thanks to the feature selection of the variance with threshold of 0.4. For the instruction filters, a maximum depth of 1 will be used. This feature will be used as reference.
- Influence of depth: This features set will be encoded as the reference, with the difference that the maximum depth will be 5. The results of this features set will allow to see the influence of the depth on the accuracy.

Table 3. Results for the different instructions selection algorithms (mean number of instruction 38,204 with out filter)

Method name	Mean extraction number after method	Hyper parameter
Line related to Android API call and constants	5048	NAN
Depth limitation	7273	1
Depth limitation	13452	10
Depth limitation	25745	50
Depth limitation +	1928	5

+ Only line related to Android API call and constants

- Typed: This features set will be encoded as the reference, however the embedding of the variables will be done on the type of variable and not on the Android array. It will show the importance of the variable content.
- Only Android: For this feature set, we will only keep the instructions related to Android API calls or Android constant. The embedding will be done as for the reference. This feature set will show the importance of Android related instructions.
- Model selection: This features set will be encoded as the reference, but the feature selection done on the Android array is done thanks to the Logistic regressions because it's the one that obtain the best result when used as a classifier.

5 Training Detail

As the objective of our research question is to see how likely LSTM is able to classify malware using decompiled source code, we have restricted our research to models Title Suppressed Due to Excessive Length 9 made up of an LSTM layer as input layer and a Dense layer with only one neuron as output layer. We used the tf.keras.optimizers.Adam optimizer and we chose to calculate the loss is binary with cross entropy function since we only have 2 class. Another constraint that we had and that the batch size could not exceed 16 because of the sizes of each sample. We implemented the model using python and keras. We can see that the input layer is indeed an LSTM layer and that this layer has input shape

```
(None, train_gene [-1]).
```

None because the model will have to manage a number of variable sequences and

```
train_gene[-1]
```

is equal to the size of a feature. We also see that we are not passing a list of features and labels to the fit method but a single object which is our generator. The models that have been shown here are the result of test and tuning of hyper parameter. Training our LSTM model has been a very time consuming task. Despite the limitations that we have imposed on ourselves, there are still a large number of hyper parameters that can be modified to increase the precision. Here is a list of the hyper parameters on which we have experimented as well as the values tested.

- Number of epochs: 16, 64, 128.
- Number of neurons in the LSTM layer: 16, 64, 128.

6 Evaluation

Table 4 shows the accuracy and the loss obtained according to the chosen feature set on the test set.

Table 4. Accuracy and the loss obtained according to the chosen feature set on the test set

Reference	Android	Model	Type	Depth	Average
Accuracy 0.775	0.828	0.838	0.734	0.744	0.7838
Loss 0.550	0.528	0.472	0.502	0.586	0.5276

We see that all the models have a good result greater than 0.7. And the average accuracy is 0.78. However, two features set have results below the average: Depth and Typed. This can be explained by the fact that the feature set of Typed does not contain information about the variable apart from the type, so if malicious methods are called, they will be ignored. However it good result seem to prove that the general structure of a malware is different from that of a benign app. For the Depth feature set, because it crawls more deeply, it has a better chance of crawling code from a trusted source package and this multiplication of "useless" instructions may drown the useful information. The two models with the best results are Android and Model. The good results of Android feature set can be explained by the fact that it has already been shown by other studies the importance of this information [11,12,17], It is therefore normal that a feature set that mainly focuses on this information gets good result. Our best accuracy is on the Model feature set. The good result of this features set can be explained by the fact that it is based on the reference feature set which already obtains good result and that the features present in variable were selected by a model which also obtained good result.

7 Conclusion

We develop a feature extraction program which decompiles an APK using Apktool and then extracts all the instructions in their most likely execution order and create different feature set depending on several methods of selection. For the purpose of answering our research question about how likely LSTM is able to classify malware using decompiled source code from the APK, we have trained a models composed of LSTM layer, to detect malware. Despite the simplicity of the model we used we were able to achieve good results showing that it is possible to detect Android malware using information extracted from the decompiled code and LSTM.

References

1. Clooke, R.: A brief history of mobile malware (2017). https://www.retaildive.com/ex/mobilecommercedaily/a-brief-history-of-mobile-malware. Accessed 19 Oct 2021
2. Statcounter: Operating system market share worldwide (2021). https://gs.statcounter.com/os-market-share. Accessed 19 Oct 2021
3. Johnson, J.: Development of new android malware worldwide from June 2016 to March 2020 (2020). https://www.statista.com/statistics/680705/global-android-malware-volume/. Accessed 19 Oct 2021
4. Zhou, Y., Jiang, X.: Dissecting android malware: characterization and evolution. In: 2012 IEEE Symposium on Security and Privacy, pp. 95–109 (2012)
5. S.R. Department: Average number of new android app releases via google play per month from March 2019 to February 2021 (2021). https://www.statista.com/statistics/1020956/android-app-releases-worldwide/. Accessed 19 Oct 2021
6. Fileinfo: .apkfile extension (2020). https://fileinfo.com/extension/apk. Accessed 19 Oct 2021
7. Tam, K., Feizollah, A., Anuar, N., Salleh, R., Cavallaro, L.: The evolution of android malware and android analysis techniques. ACM Comput. Surv. **49**, 1–41 (2017)
8. Milosevic, N., Dehghantanha, A., Choo, K.-K.R.: Machine learning aided android malware classification. Comput. Electr. Eng. **61**, 266–274 (2017) https://www.sciencedirect.com/science/article/pii/S0045790617303087

9. Almin, S.B., Chatterjee, M.: A novel approach to detect android malware. Procedia Comput. Sci. **45**, 407–417 (2015). International Conference on Advanced Computing Technologies and Applications (ICACTA). https://www.sciencedirect.com/science/article/pii/S1877050915004135
10. Talha, K.A., Alper, D.I., Aydin, C.: Apk auditor: permission-based android malware detection system. Digit. Invest. **13**, 1–14 (2015). https://www.sciencedirect.com/science/article/pii/S174228761500002X
11. Wu, D., Mao, C., Wei, T., Lee, H., Wu, K.: Droidmat: android malware detection through manifest and API calls tracing. In: Seventh Asia Joint Conference on Information Security 2012, pp. 62–69 (2012)
12. Aafer, Y., Du, W., Yin, H.: DroidAPIMiner: mining API-level features for robust malware detection in android. In: Zia, T., Zomaya, A., Varadharajan, V., Mao, M. (eds.) SecureComm 2013. LNICST, vol. 127, pp. 86–103. Springer, Cham (2013). https://doi.org/10.1007/978-3-319-04283-1_6
13. Android developers. Common intents. https://developer.android.com/guide/components/intents-common#java. Accessed 19 Oct 2021
14. Android developers: monkeyrunner (2020). https://developer.android.com/studio/test/monkeyrunner. Accessed 19 Oct 2021
15. VirusShare: Virusshare.com (2020). https://virusshare.com/. Accessed 19 Oct 2021
16. Allix, K., Bissyandé, T.F., Klein, J., Le Traon, Y.: Androzoo: collecting millions of android apps for the research community. In: Proceedings of the 13th International Conference on Mining Software Repositories, MSR '16, pp. 468–471. ACM, New York (2016). http://doi.acm.org/10.1145/2901739.2903508
17. Arp, D., Spreitzenbarth, M., Hubner, M., Gascon, H., Rieck, K., Siemens, C.: Drebin: effective and explainable detection of android malware in your pocket. In: NDSS, vol. 14, pp. 23–26 (2014)

Check for updates

Vulnerability Detection Using Deep Learning

Mahmoud Osama Elsheikh(✉)

Heriot-Watt University, Edinburgh, UK
moe2@hw.ac.uk

Abstract. Software vulnerabilities are one of the main entry points of cyberattacks. Several papers focused on the challenge of automatically identifying vulnerabilities and recent works started applying machine learning techniques. With the advent of newer deep learning technologies like transformers, there is more room for improvement in vulnerability detection tools. This survey explores the various techniques used to identify vulnerabilities in code, from recurrent neural networks to transformers, such as the Bidirectional Encoder Representations from Transformers (BERT).

1 Introduction

A computer executes code exactly as it is written. While its execution is flawless, the underlying code is written by humans, and as a result contains mistakes. These mistakes while writing code can lead to flaws/bugs that can be exploited by a malicious third party. A flaw that can be exploited is a security vulnerability. Vulnerabilities are assigned severity scores [3], where higher scores are given to vulnerabilities with more critical consequences.

Flaw detection is as old as code writing itself. It consists of detecting as many software flaws as feasible. The earliest methods of detection in code could be categorised under two categories, namely static and dynamic analysis [24]. Static analysis tools [7] scan the source code for flaws, and are intended to identify common pitfalls such as buffer overflows. The naming comes from the fact that static analysis tools do not run the code and only identify errors by checking the code itself, as opposed to dynamic analysis where vulnerabilities are identified by running sections of the code and checking for any issues that arise during runtime [7].

Examples of static analysis tools include Flawfinder [23] and Yasca [21]. These tools work by comparing the functions in the source code to a pre-existing database of known functions that can cause problems, such as the C function sprintf(), which can cause buffer overflow issues. The identified functions are then tagged and put into a list, sorted by risk, which is then presented to the programmer in order to fix the potential flaws.

However, as mentioned in evaluations of static analysis tools, they have several limitations [16]. They can create a high rate of false positives, flagging code

© The Author(s), under exclusive license to Springer Nature Switzerland AG 2022
H. Ragab Hassen and H. Batatia (Eds.): ACS 2021, LNNS 378, pp. 53–60, 2022.
https://doi.org/10.1007/978-3-030-95918-0_6

fragments that are not malicious. These tools could also fail to identify certain vulnerabilities, resulting in false negatives.

Dynamic analysis heavily depends on software tests. This can be time consuming and an insufficient amount of tests could lower the confidence level in the vulnerability detection.

Both classic static and dynamic approaches use predefined databases of vulnerabilities to check against. The use of predefined databases is a valid approach, but comes with limitations. The scope of the vulnerabilities that such a system can find are limited to what it has previously identified. To go beyond such a limitation, the community adopted more adaptive and less rigid machine learning techniques, inspired by text analytics [13, 15].

However, machine learning is not inherently catered to vulnerability detection, so a first challenge is in adapting the vulnerability detection problem to formulate as a machine learning, or text-analytics, one. These techniques were created to process natural languages, and their application for code processing requires particular adaptation. Indeed, when dealing with code analysis, code chunks are not self-contained. Indeed, a function call has references to other parts of the software where it is called, variables could be used across multiple lines of code, etc.

The rest of this proposal is organised as follows: Sect. 2 reviews important concepts, notable related work and identifies gaps in the literature and Sect. 3 discusses future work that could be done.

2 Background

Natural language processing is suited for vulnerability detection because of the similarities found between code and natural language. Fundamentally, code is a sequence of words, allowing researchers to break down a complex function into a series of words. This series can then be manipulated using various techniques.

2.1 Recurrent Neural Networks

Recurrent Neural Networks (RNN) introduce a concept of "memory" which allows it to retain more information when presented with longer sequences of text [20]. This has proven more beneficial over standard feedforward neural networks, but suffered from its own downfalls. As the sequence got longer, the gradient used for improving the parameters moves towards zero, leading to a vanishing gradient problem. Vanishing gradients make it difficult for the model to continue being improved, as any new changes to the gradient are too small to be significant.

The Long Short Term Memory (LSTM) model is an RNN variant that addresses the vanishing gradient problem by introducing "forget gates". Forget gates allow the cells to retain less information about previous words in the chain. This is useful because while the words might be related to one another over a small to medium distance, as the distance between the words increases, the less related they are to each other [10].

For vulnerability detection, one of the earliest papers to incorporate machine learning was VulDeePecker [13]. The dataset was based on the National Institute of Standards and Technology's "NVD" database [4] combined with the Software Assurance Reference Dataset (SARD) [5]. These datasets contained vulnerable C/C++ code from various repositories, marked as either vulnerable or not vulnerable. Instead of being directly used, the dataset was transformed into a custom representation called a "code gadget". Code gadgets were smaller than the original datasets and had a different structure. Using Checkmarx [1], Li et al. extracted library and API function calls. These calls are combined with any semantically relevant lines to create a code gadget. From this, 61,638 samples were created. LSTM models only take vectors as input, which meant that code gadgets would need to be further processed. In order to handle this, the text code gadgets were converted with word2vec [17] into a vector representation.

The neural network model itself is split into three layers. An initial LSTM layer first processes the input vector. Then, a dense layer takes in the high dimensional output of the LSTM layer and reduces the number of dimensions. This leads to an improvement in processing time and can also yield improvements in the final result, as it lowers overfitting. Finally, the low dimension vectors are sent into a softmax layer, which is responsible for turning these vectors into a single probability distribution. The probability distribution indicates the chance of the given code gadget containing a vulnerability. This result is also used to perform the training process. In training, the parameters of the layers are tweaked to yield better results by comparing the final softmax output to the true label given in the datasets.

Instead of identifying vulnerabilities in the source code, Lu's paper [15] identifies whether a given executable is vulnerable or not. The dataset is comprised of 1092 samples, 969 vulnerable and 123 not vulnerable samples. These were created from executables collected from three sources: hand-collected executables, malware from Virusshare [6] and MalwareDB [2], and Microsoft's Malware Classification Challenge [19]. Each executable is disassembled by using IDA pro to be converted to assembly, then converted into an opcode sequence by using a custom algorithm. Attached to every sample is a label, identifying what class of vulnerability it is: trojan, worm, adware, backdoor, downloader and not vulnerable.

Since opcode sequences are strings, they cannot be parsed by the LSTM model. To convert them to vectors, one-hot encoding is used to create an encoding based on the frequency of opcodes. This representation is 391 dimensions in size, so it is cut down to 100 dimensions. Unfortunately, one-hot encoding does not convey semantic relations between the words, so the one-hot representation is then encoded using the continuous bag of words model.

The LSTM model itself is comprised of four layers: 2 LSTM hidden layers, a mean-pooling layer, and a softmax output layer. The LSTM layer receives the vector representation and then converts it to a feature representation. While the traditional output of an LSTM is only the final feature representation, the mean-pooling layer can iterate over all LSTM outputs and create an average. This average improves variance. Lastly, the softmax layer converts the averaged feature representation into the likelihood of the sample being any given category.

2.2 Convolutional Neural Networks

Convolutional Neural Networks (CNN) were originally used for computer vision. It is capable of processing spatial data, which works well for classifying objects in images. The architecture of this algorithm involves relying on multiple convolutional and pooling layers, which slide over the dataset and reduce the dimensionality of the data. Because this approach has a singular window that slides over a specific chunk and then moves left or right, it can easily identify spatial trends. The sliding window approach is also applicable to text data, as viewing a group of words can better explain their meaning over viewing the words individually. Seeing this, CNNs were applied to NLP problems such as sentiment analysis and text classification problems [11].

CNNs were used for code analysis. Wu et al. trained a convolutional neural network in order to identify vulnerabilities inside Linux binary programs [25]. The dataset is created from 9872 binary programs in "/usr/sbin" and "/src/bin". These binaries are executed and their events are hooked using pythontrace binding and collected into a sequence. These events are made of C function calls. Included with the function calls was the final state of the binary program's execution. The labels were created by assessing the program's final state with a fuzzer. If the program exits successfully, then the sample is not vulnerable, but if there was a crash, then the sample is labelled vulnerable.

From there, the dataset itself would enter pre-processing. First, the data is tokenized and then the length of each item is controlled. This is done as the neural network requires fixed length input. Tokens were randomly cut for samples that were too long. For samples that were too short, padding with zeros would occur. These samples would then be converted into a vector.

The neural network itself is made of multiple layers. The input itself is a sentence with word embeddings, this is then passed through multiple convolutional layers and dense layers to be processed. As the dataset traverses the layers, it also passes through a max pooling layer that helps lower the dimensionality even further. Throughout the process, the results remain positive through the use of rectified linear units or ReLU. These layers are not classifying, so the final layer is a sigmoid layer that outputs a final classification. Compared to LSTMs, the spacial structure of CNNs allow them to better learn patterns in data. This also means they are faster than LSTMs. However, they are not capable of capturing all context due to their approach.

Another approach is VulExplore. [9] In that paper, Guo et al. detect vulnerabilities in source code using a CNN by analyzing code metrics. Their dataset base is a public dataset of labelled code slices made by Li et al. [12]. This data comprises of over 420,000 samples spread across 4 categories: Improper library and API calls, improper use of arrays, improper use of pointers, and integer overflows and other arithmetic errors.

In order to ease the computational load, a smaller sample size focusing on only pointer errors was chosen. Each sample was then passed through a series of equations in order to calculate the code metrics such as program vocabulary and cyclomatic complexity. The 8 chosen metrics were used in order to create a code metric database for each sample. In total, the final dataset contained 65,513 samples, 9,314 of which are vulnerable. These samples were encoded as a three dimensional vector which contained the code metrics. Alongside this, a binary label was used to indicate if the sample is vulnerable (1) or not vulnerable (0).

The model itself was consistent of 4 layers. The CNN layer extracts feature vectors from the input vector. The LSTM layer creates a deeper representation by taking the feature vectors from the CNN layer and helping to map whether a given feature leads to vulnerabilities or not. The vector output of the LSTM layer is too high dimension, so it is sent to a dense layer for dimensionality reduction and finally a softmax layer for classification.

2.3 Transformers

As research progressed on the various techniques of natural language processing, one of the newer additions to RNN based systems was the attention mechanism. Attention aimed to solve the problem of maintaining prior context. Maintaining a small amount of information from each prior word would lead to the vanishing gradient problem, where the changes become so small that they provide a negligible impact on the actual results. Attention systems bypassed this problem by allowing the model to access the prior state as usual and a global state that has all the words in the given sentence. However, it was soon discovered that relying on attention alone would yield results that surpassed RNN models, as was mentioned in [22]. This discovery led to the creation of the transformer model, a model that did not rely on any convolution or recurrent mechanisms. It would only use attention in an encoder-decoder structure.

Another development was the use of transfer learning. This is where models trained on a specific dataset can apply what they learned on a new context. This technology developed into the practice of pre-training, where a given model would be trained on a particular dataset and then packaged to be used in other contexts.

One of these models is the Bidirectional Encoder Representations from Transformers (BERT) [8]. This pre-trained model is trained by using english wikipedia as a baseline. It performed well in various fields of natural language processing, such as machine translation and sentiment analysis.

With the context of sentiment analysis, the learned knowledge of the pre-trained model is based on the english language and is then applied onto the english language. In translation, the transfer learning extends to finding patterns in foreign languages. Researchers then set their sights on whether or not transfer learning would work on code text. One of those papers was the work of Ziems and Wu, who used BERT for vulnerability detection [26]. Their goal was to classify a given code sample into The model being primarily tested was BERT, with LSTM being used as a baseline. The BERT model's output would be a probability distribution, giving a score for the likelihood of any given dataset item to be a given class. The class with the highest score would then be the classified class for that dataset item. This class can either be "Not Vulnerable" or one of 132 different vulnerabilities.

To start, their dataset was the Software Assurance Reference Dataset (SARD) which contained examples of vulnerable C and C++ code. The functions were stripped of excess information like the labels "good" and "bad" on the function names and comments. This would convert a function name like "CWE121_Buffer _Overflow_badSink" to "func1". The cleaned functions are then passed through a tokenizer and then embedded into a vector.

The BERT model's initial implementation was to split the dataset files into 256 word sequences, due to the limitations that BERT had. This model was effective, but contained context issues. The model could only recognize the given 256 word sequence, and would not remember anything before or after. However, even at this naive state, the model outperformed LSTMs.

There was room for improvement. This came in the form of combining the output of the BERT model for any given sequence with an RNN model, like a uni or bi-directional LSTM. This combination led to the state of the art results produced by this paper. The success is attributed to the LSTM being able to retain more context than the BERT model, while leveraging BERT's performance improvements on a sequence to sequence level.

2.4 Gap Identification

The vulnerability detection problem can be split into two sections: the dataset and the machine learning model. For any improvements to occur in this field, either one must be improved in order to yield better results.

Dataset problems stem from the fact that a codebase can be massive, with many thousands of lines and varying locations of vulnerabilities. From the start, finding vulnerabilities to analyze can be difficult. Current approaches to this problem involve collecting samples from pre-existing databases such as the SARD. For a custom codebase, vulnerability identification and labeling must be done either by hand using human experts or relying on pre-existing commercial analysis tools. However, there also exists another way of collecting data from open source repositories. This approach relies on the difference between bugfix commits. This will essentially create two data items, one with the vulnerability and one without. This unfortunately comes with its own set of issues, such as tracking down such bugfix commits and making sure that the changes were actually addressing the issue specifically, and not adding other features.

Furthermore, once the code samples are obtained, there is the issue of organizing them. In order to fit the format of machine learning models, the code data must be converted into a vector. This vectorizing can have many decision points, such as picking the tokenizer or picking the specific vector representation.

The other improvement point is in the machine learning model itself. There are many machine learning models to pick from, from LSTM to CNN to newer Transformer models. Each model comes with its own unique set of advantages and issues. Furthermore, as natural language processing evolve, newer variants of models are continuously being proposed. These models could yield performance improvements over the current state of the art. Furthermore, combining the models could potentially lead to better results. This can be seen in [26], where the optimal model combined a BERT model with a BILSTM one.

Pre-training has become standard practice with newer transformer-based models like BERT. Thanks to the similarities between code and the English language, English pre-trained models, such as those pre-trained on Wikipedia, can be applied on code However, there might be a potential for improvement if the model was to be pre-trained on code-based data.

3 Future Work

There is plenty of room for future improvements. For the dataset aspect, collecting more data is always possible. Furthermore, the locations that data can be collected from vary. One such example is [18], where the data is sourced from open source commits that had previously patched a vulnerability. For the model aspect, one of the major areas of development is the transformer model. Models such as BERT are pre-trained on general data, so investigating the effects of domain-specific data has the potential to improve the model's performance.

Furthermore, using different BERT variants can lead to an improvement in results. Examples of variants include simple ones like BERT-Large, which simply contains double the transformer layers, to entirely new models like RoBERTa [14].

Lastly, BERT is not infallible, and some of the fundamental flaws it contains have potential to be resolved by combining it with other machine learning techniques such as LSTMs or graph convolutional networks.

References

1. checkmarx. https://checkmarx.com
2. Malwaredb. http://malwaredb.malekal.com
3. Nist: The common vulnerability scoring system. https://nvd.nist.gov/vuln-metrics/cvss
4. Nvd: National vulnerability database. https://nvd.nist.gov
5. Sard: Software assurance reference dataset. https://samate.nist.gov/SARD/index.php
6. Virusshare. https://virusshare.com
7. Aggarwal, A., Jalote, P.: Integrating static and dynamic analysis for detecting vulnerabilities. In: 30th Annual International Computer Software and Applications Conference (COMPSAC'06), vol. 1, pp. 343–350 (2006). https://doi.org/10.1109/COMPSAC.2006.55

8. Devlin, J., Chang, M.W., Lee, K., Toutanova, K.: Bert: Pre-training of deep bidirectional transformers for language understanding (2019)
9. Guo, J., Wang, Z., Li, H., Xue, Y.: Detecting vulnerability in source code using CNN and LSTM network. Soft Comput. (2021). https://doi.org/10.1007/s00500-021-05994-w
10. Hochreiter, S., Schmidhuber, J.: Long short-term memory. Neural Comput. **9**(8), 1735–1780 (1997). https://doi.org/10.1162/neco.1997.9.8.1735
11. Kim, Y.: Convolutional neural networks for sentence classification (2014)
12. Li, Z., Zou, D., Xu, S., Jin, H., Zhu, Y., Chen, Z.: SySeVR: a framework for using deep learning to detect software vulnerabilities. IEEE Trans. Dependable Secure Comput. 1 (2021). https://doi.org/10.1109/TDSC.2021.3051525
13. Li, Z., et al.: VulDeePecker: a deep learning-based system for vulnerability detection. In: Proceedings 2018 Network and Distributed System Security Symposium (2018). https://doi.org/10.14722/ndss.2018.23158
14. Liu, Y., et al.: Roberta: a robustly optimized BERT pretraining approach (2019)
15. Lu, R.: Malware detection with LSTM using opcode language (2019)
16. Mahmood, R., Mahmoud, Q.H.: Evaluation of static analysis tools for finding vulnerabilities in java and c/c++ source code (2018)
17. Mikolov, T., Chen, K., Corrado, G., Dean, J.: Efficient estimation of word representations in vector space (2013)
18. Perl, H., et al.: VCCFinder: finding potential vulnerabilities in open-source projects to assist code audits. In: Proceedings of the 22nd ACM SIGSAC Conference on Computer and Communications Security, CCS '15, pp. 426–437. Association for Computing Machinery, New York (2015). https://doi.org/10.1145/2810103.2813604
19. Ronen, R., Radu, M., Feuerstein, C., Yom-Tov, E., Ahmadi, M.: Microsoft malware classification challenge (2018)
20. Rumelhart, D.E., Hinton, G.E., Williams, R.J.: Learning representations by back-propagating errors. Nature **323**(6088), 533–536 (1986). https://doi.org/10.1038/323533a0
21. Scovetta, M.: YASCA: yet another source code analyzer. https://github.com/scovetta/yasca
22. Vaswani, A., et al.: Attention is all you need (2017)
23. Wheeler, D.A.: Flawfinder. https://dwheeler.com/flawfinder/
24. Wichmann, B., Canning, A., Marsh, D., Clutterbuck, D., Winsborrow, L., Ward, N.: Industrial perspective on static analysis. Softw. Eng. J. **10**(2), 69 (1995). https://doi.org/10.1049/sej.1995.0010
25. Wu, F., Wang, J., Liu, J., Wang, W.: Vulnerability detection with deep learning. In: 2017 3rd IEEE International Conference on Computer and Communications (ICCC), pp. 1298–1302 (2017). https://doi.org/10.1109/CompComm.2017.8322752
26. Ziems, N., Wu, S.: Security vulnerability detection using deep learning natural language processing (2021)

Feature Selection Approach for Phishing Detection Based on Machine Learning

Yi Wei[1(✉)] and Yuji Sekiya[2]

[1] Graduate School of Engineering, The University of Tokyo, Tokyo, Japan
weiyi@g.ecc.u-tokyo.ac.jp
[2] Graduate School of Information Science and Technology, The University of Tokyo, Tokyo, Japan
sekiya@nc.u-tokyo.ac.jp

Abstract. Phishing is a kind of cybercrime that uses disguised websites to trick people into providing personally sensitive information. Phishing detection with high accuracy has attracted enormous interest in Cyber Security. Many website-based features have been applied for phishing detection, however, useless features can lead to extra feature extraction and phishing detection time costs. This paper analyzes 111 features of the latest published phishing websites dataset to investigate the obvious differences and correlations between phishing and legitimate websites. By applying eleven commonly used Machine Learning algorithms and evaluating their performances, we choose Random Forest as our phishing detection algorithm. Based on feature importance methods, we proposed a framework to reduce the number of features while maintaining high detection accuracy. The model training time and the usage of memory can also be saved. With the combination of the proposed feature selection framework, the important findings indicate that by using only 14 features, the accuracy of phishing detection can achieve 97.0%.

Keywords: Cyber security · Machine learning · Phishing website detection · Feature selection

1 Introduction

Global cyber threats continue to evolve at a rapid speed, causing a rising number of data breaches every year. Malicious criminals are responsible for most incidents of data breaches. Phishing is a kind of cybercrime, which involves luring the user into providing sensitive and confidential information to the attacker. The information includes address, credit card details, bank account details, passwords for online shopping sites, and other personally sensitive information. Phishing of private information on the web has caused havoc on a majority of users due to the lack of internet security awareness.

Due to the COVID-19 pandemic in the world, great changes have taken place in people's lifestyles. People around the world are becoming more and more accustomed to shopping, learning, and remote working online. Many coronavirus-related apps about

© The Author(s), under exclusive license to Springer Nature Switzerland AG 2022
H. Ragab Hassen and H. Batatia (Eds.): ACS 2021, LNNS 378, pp. 61–70, 2022.
https://doi.org/10.1007/978-3-030-95918-0_7

testing, treatments, cures, and remote works were developed and used widely. The widespread use of the Internet makes phishing even more offensive. According to the Phishing Activity Trends Report, 1st Quarter 2021 published on 8 June 2021 from the Anti-Phishing Working Group (APWG) [1], The number of phishing attacks observed by the APWG and its contributing members doubled over the course of 2020. Attacks then peaked in January 2021, with an all-time high of 245,771 new phishing sites appearing in that month alone [2].

Detection of phishing attacks with high accuracy has always been an issue of great interest. In the case of the development and the popularity of Artificial Intelligence, there is an increasing number of phishing detection approaches to improve the classification and downsize the labor cost. This research applies various advanced Machine Learning technology to the latest published phishing dataset, aiming at finding the best suitable algorithm for phishing detection. Based on the observation of differences between legitimate and phishing websites and the suitable Machine Learning Algorithm, the ultimate goal is to select as few efficient features as possible while maintaining high detection accuracy and low memory usage.

This paper follows the following structure: Sect. 2 introduces the related works in phishing detection based on Machine Learning, Sect. 3 is the description and analysis results of phishing websites based on the latest dataset, Sect. 4 shows the experimental detection results by using various Machine Learning algorithms. Section 5 are the proposed feature selection approaches and selection results. Section 6 presents the conclusion and future works.

2 Related Works

While there are many proposals to detect phishing websites. These methods are to be revised from time to time due to the innovativeness of new phishing attacks. Most of the machine learning algorithms used in phishing detection are categorized as supervised machine learning. That is where a classifier tries to learn certain characteristics of various phishing and legitimate websites to predict a response. Machine learning based on anti-phishing solutions extract features like URLs [3], hyperlinks information [4], page content [5], digital certificate, website traffic, and other resources. The accuracy of the anti-phishing solution depends on features set, training data, and machine learning algorithms.

By analyzing the URL of the webpage, M. Korkmaz et al. determine 58 features from URLs and compare 8 different algorithms [6]. As result, the RF classifier shows the highest accuracy rate of 94.59%, 90.50%, and 91.26% in three datasets. It also shows that ML algorithms are efficient with the change of datasets. Based on the observation of a number of phishing websites are imitating common websites to deceive web users, Wu et al. use Selenium [7] as the crawler to get real-time content on the web to find the URL hidden by the web designer in JavaScript rendering. By using SVM, the accuracy of detecting phishing webpages is 89.3%, while the false positive rate is 6.2%. Lokesh et al. use many techniques [8] such as Decision tree Classifier, K nearest neighbours, Linear SVC classifier, Random Forest classifier, One-class SVM classifier and out of which they observed that Random Forest got the highest accuracy of about

96.87%. Researchers in Cornell University [9] use 11,000 sample websites from PhishTank and 30 features extracted themselves. They evaluate twelve classifiers to compare the results, they get very good performance in ensemble classifiers namely, Random Forest, XGBoost both on computation duration and accuracy, which shows ensemble-based learning algorithms can combine several weak learners into a stronger one. A review of ML-based approaches for phishing detection is conducted in a survey paper [10], which presents a comprehensive review of conventional ML techniques that are significant for the detection of malicious attacks on websites.

The datasets used are always provided by UCI and Mendeley or extracted by themselves in researchers' works. However, different kinds of features result in dataset redundancy because some features may have a high correlation. Besides, useless features may introduce noise and have a negative effect on classification. Therefore, feature selection work is necessary to further select the most significant features to reduce feature redundancy and enhance the detection performance.

K.L. Chiew et al. propose a new feature selection framework [11] for machine learning-based phishing detection system, called the Hybrid Ensemble Feature Selection (HEFS), where existing filter measures are leveraged to find an effective subset of features for utilization. As part of the HEFS, they propose a novel algorithm called CDF-g to automatically determine the optimal number of features. S. Shabudin et al. present the performance of two feature selection techniques known as the Feature Selection by Omitting Redundant Features (FSOR) and Feature Selection by Filtering Method (FSFM) [12]. The results demonstrate that the FSOR method is statistically significant and outperforms the other method when using Random Forest classifiers.

3 Phishing Websites Feature Analysis

The most challenging problem in phishing detection is the dataset. If there is a worldwide accepted dataset, researchers can compare their models' performances convectively. But because the phishing website has the characteristics of short live time and fast update. There are not standard feature set to record the feature of phishing websites. Many researchers implemented code to extract features from PhishTank [13], which is a collaborative clearinghouse for data and information about phishing on the Internet.

The dataset used in our experiment is from [14], Grega Vrbančič et al. collected and presented two dataset variations in October 2020, which consist of 58,645 and 88,647 websites labeled as legitimate or phishing, respectively. The list of legitimate URLs was obtained from Alexa ranking website and the phishing websites were also obtained from PhishTank. The dataset is based on the uniform resource locator (URL) properties, URL resolving metrics, and external services. It includes in total features 111 attributes excluding the target phishing attribute, which denotes whether the particular instance is legitimate (value 0) or phishing (value 1). The dataset can be obtained from Mendeley Data [15].

Several experiments have been carried out to analyze the 111 features in dataset Mendeley_2020. Figure 1 shows the basic structure of a URL, the contributors of the dataset Mendeley_2020 extracted features by splitting the URL into 5 parts including Domain, Directory, File, Parameters and the whole URL. Then they counted the quantity

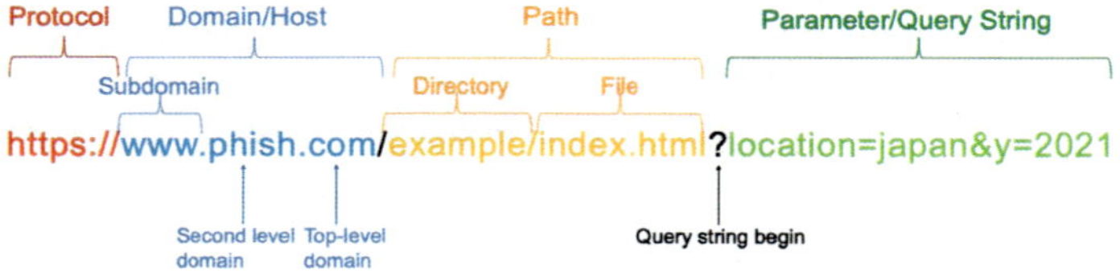

Fig. 1. An example of URL structure

of 17 kinds of signs in each part as features. Figure 2 presents the average number of signs in each part. We can see obviously that in phishing websites, there are a greater number of dot (.), hyphen (-), underline (_), slash (/), questionmark (?), equal (=), at (@), and (&) in the whole URL. Nevertheless, in legitimate websites, more number of percent (%) shows in file and parameters.

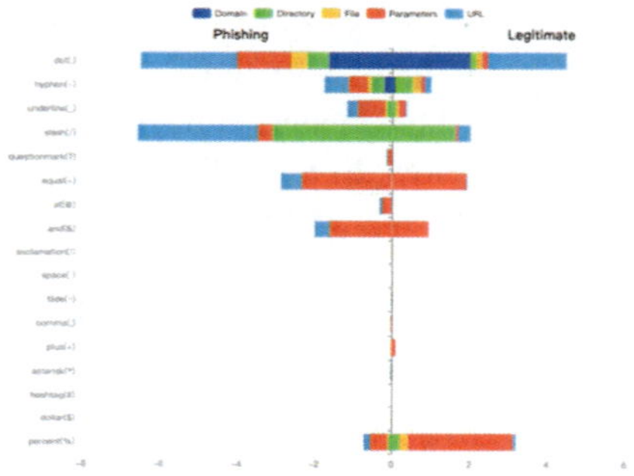

Fig. 2. Average number of signs in each part of URL

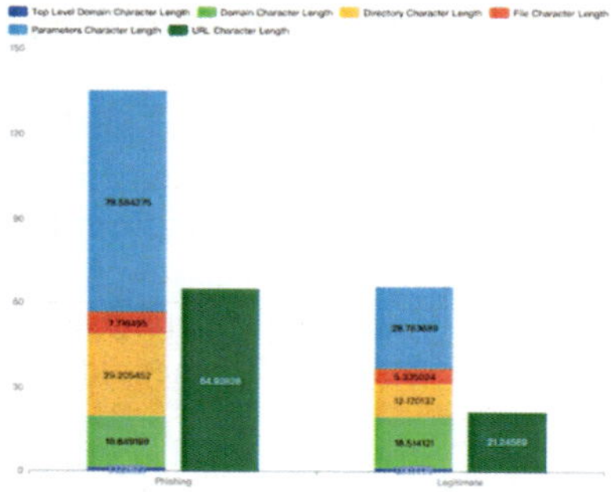

Fig. 3. Average length of each part in URL

Beside signs in URL, the length of each part and the total URL length can help to identify phishing websites. Figure 3 illustrates the average length of different part in URL. The average number of characters in whole URL of phishing websites is 3 times as much as that of legitimate websites. As for directory, it's 2.5 times, for parameters, it's 2.8 times. In general, phishing websites usually have longer URL.

4 Phishing Detection Based on ML

Machine learning models are being extensively used by leading internet service providers like Yahoo, Gmail, and Outlook, to filter and classify [16]. We applied eleven machine learning algorithms for phishing website detection including Logistic Regression, Linear Discriminant Analysis, Classification and Regression Tree, Support Vector Machine, Naive Bayes Classifier, K-Nearest Neighbor, Random Forest, AdaBoost, GBDT, XGBoost, and LightGBM.

The models created with these algorithms are trained by using the scikit-learn (0.24.2) library in the Python (3.8.3) programming language in Jupyter Notebook (6.0.3). In our experiments, we applied 10-fold cross-validation 10 times to the dataset Mendeley_2020_small. Table 1 is the classification results (Fig. 4).

Table 1. Classification results by using conventional ML algorithms

Classifier	Accuracy	Precision	Recall	F1 score
LR	0.8539	0.8644	0.8553	0.8593
LDA	0.8777	0.8408	0.9453	0.8899
CART	0.9284	0.9287	0.9329	0.9317
SVM	0.7227	0.7031	0.8130	0.7540
NB	0.7288	0.9176	0.5295	0.6712
KNN	0.8375	0.8366	0.8569	0.8466
RF	0.9547	0.9525	0.9603	0.9567

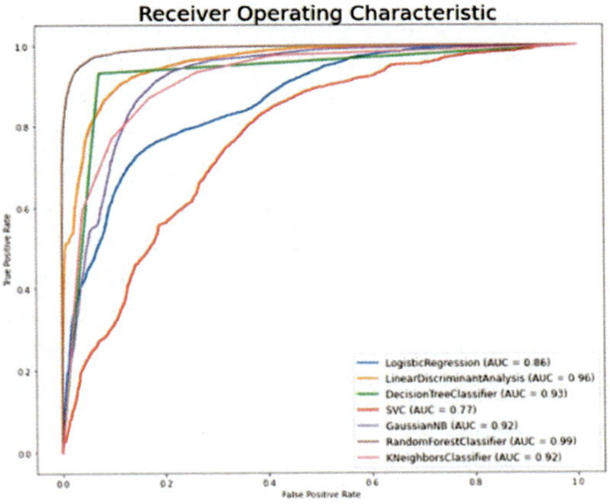

Fig. 4. ROC of 7 conventional ML prediction methods

From the result, we found that Random Forest is highly accurate than other 6 algorithms. Whereas SVM and NB have a lower accuracy rate. The highest AUC of RF in ROC shows that the RF classifier is able to distinguish between positive class and negative class correctly. The results above represent that Random Forest Classifier is suitable for detecting phishing websites. In the same way, other ensemble methods are compared in Table 2.

XGBoost and LightGBM have not been integrated with Sklearn, extra packages need to be installed. The version of XGBoost used in the experiment is 1.3.3 and the version of LightGBM used is 3.2.1. LightGBM shows the fastest training time with 0.7782 s for 10000 datasets, it also shows the best performance in predicting accuracy. Given the

Table 2. Classification results based on 5 Ensemble ML methods

Classifier	Accuracy	Precision	Recall	F1 score	Training time (s/w)	Testing time (s/w)
RF	0.9547	0.9525	0.9603	0.9567	6.5441	0.2540
AdaBoost	0.9323	0.9343	0.9363	0.9353	9.9455	0.9730
GBDT	0.9593	0.9603	0.9620	0.9611	38.4067	0.1864
XGBoost	0.9591	0.9593	0.9627	0.9610	14.5773	0.0491
LightGBM	0.9617	0.9628	0.9639	0.9634	0.7882	0.1234

above, the Ensemble Machine Learning Methods represented by Random Forest have better performances in detecting phishing websites. For this reason, we choose Random Forest as the basic detecting algorithm in our feature selection approach.

5 Feature Selection Approach

5.1 Approach Description

Explainable Artificial Intelligence is an emerging research direction helping the user or developer of machine learning models understand why models behave the way they do. The most popular explanation technique is feature importance [17]. The feature importance describes which input features are relevant and how useful they are at predicting the results. There are many different methods to measure feature importance including MDI (Mean Decrease in Impurity), Permutation Feature Importance and SHAP (SHapley Additive exPlanation) interpretation. Through experiments, we found that the most important features differ depending on the methods used, by using a combination of several feature importance methods could provide more reliable and trustworthy results. The flow chart in Fig. 5 shows the process of feature selection framework we proposed.

Data preprocessing changes the original features into a form more suitable for machine learning models. Then, by using Variance Threshold, features with zero variance are filtered and removed. Next, all the three feature importance methods mentioned

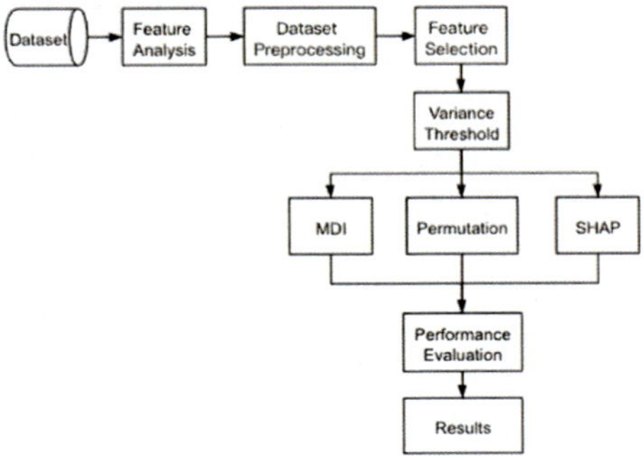

Fig. 5. Proposed feature selection approach

before will be used to get a feature importance ranking. The final sorting will be the best sorting manually selected from the three sorting results. With this ranking list, we might think "how many features will be enough for phishing detection?", we need a balance to reduce the number of features while maintaining a relatively high accuracy.

As shown in Fig. 6, here, we define an "Anti-Phishing Score", which is the weighted average of four standardized evaluation metrics: accuracy, recall, testing time and memory usage. We use the reciprocal of testing time and memory usage and normalize the four metrics range from 0 to 1 by using MinMaxScaler, which can transform features individually by scaling each feature to a given range. Then the "Anti-Phishing Score" can be calculated. The most appropriate set of features will be selected in order of the Anti-Phishing Score.

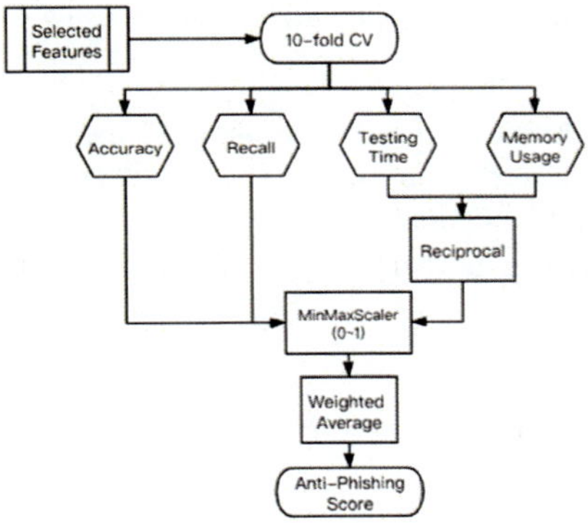

Fig. 6. Performance evaluation process of the selection approach

5.2 Experimental Results

The dataset used in this experiment is Mendeley_2020_full which has 111 features and 88647 instances. The accuracy of the detection by using feature subsets with different number of features is in Fig. 7: if we just use the feature "directory _length" to train RF model, the detection accuracy can achieve 89.4228%; if we add one more feature "time_domain_activation", the detection accuracy will raise to 92.4685%, then add feature "asn_ip" in the subsets, the accuracy will be 95.5330%. If we use only one method alone, we cannot get the optimal result of sorting. Combination will offer more trustworthy solutions.

MDI		Permutation		SHAP Values	
Number of Features	Accuracy (%)	Number of Features	Accuracy (%)	Number of Features	Accuracy (%)
1 (directory_length)	**89.4228**	1 (time_domain_activation)	75.4164	1(time_domain_activation)	75.4164
2 (+time_domain_ac)	**92.4685**	2 (+directory_length)	92.4685	2 (+directory_length)	92.4685
3 (+qty_?_directory)	92.4121	3 (+asn_ip)	**95.5330**	3 (+length_url)	93.4349
4 (+length_url)	93.5025	4 (+qty_._domain)	**95.7736**	4 (+qty_._domain)	93.9914
5 (+qty_/_url)	93.9387	5 (+ttl_hostname)	**96.1835**	5 (+qty_?_directory)	93.9462
10	96.3377	10	96.9656	10	95.6308
15	96.5708	15	97.1423	15	96.5896
20	96.7212	20	97.1085	20	96.6949
25	96.6422	25	97.1874	25	96.6761
30	96.8302	30	97.1799	30	97.1536
35	97.0822	35	97.1724	35	97.0746
40	97.1160	40	97.2062	40	97.0295

Fig. 7. Accuracy with number of features using 3 methods

The proportion of each metric can be adjusted to meet different requirements, for example, If the memory is not taken into consideration, the memory metric can be deleted, if one detection device focuses more on detecting time cost, the weight of the testing time metric can be higher. We use the weight "2:2:1:1" to calculate the Anti-Phishing Score in our experiment.

Table 3. Feature selection results

Ranking	Anti-phishing score	Num. of features	Accuracy	Recall	Testing time (s)	Memory (MB)
1st	0.71006	14	0.97033	0.96221	0.27865	10.1
2nd	0.70787	10	0.96928	0.96140	0.28844	7.4
3rd	0.70286	26	0.97185	0.96270	0.29210	18.3
4th	0.70271	11	0.96964	0.96153	0.30168	8.1
5th	0.70254	15	0.97067	0.96208	0.29884	10.8

Table 3 shows that 14 features will be enough for detecting phishing website by following the balanced metric "Anti-Phishing Score", it can reduce 86.66% usage of memory while maintaining 97% (suffer from a minimal accuracy deterioration of 0.173%) detection accuracy. The selected feature sets can be shown in Table 4.

To further validate the effectiveness of the proposed approach, we also applied our feature selection approach to dataset Mendeley_2018 [18] (which has 10000 instances and 48 features) to check the results, our framework can select 23 features and the accuracy can be 97.78%. Compared with [11], who selected 10 features from Mendeley_2018 as a baseline showed 94.6% accuracy, by using the top-10 features we selected, accuracy can achieve 96.83%. When they use full 48 features, the detection accuracy is 96.17%, our detection reaches peak 98% when we use 35 features (Table 5).

Table 4. Selected features for phishing detection

1 directory length	2 time_domain_activation	3 Autonomous System Number
4 number of dot in domain	5 ttl_hostname (Time-To-Live)	6 url length
7 time_domain_expiration	8 num of Mail eXchanger Servers	9 domain lookup time response
10 num of resolved Name Server	11 number of dot in url	12 number of hyphen in url
13 file length	14 number of slash in url	

Table 5. Performance comparison between proposed framework and HEFS

Feature selection method	Feature set	Number of features	Accuracy (%)
HEFS [11]	Full	48	96.17
HEFS [11]	Baseline	10	94.60
Proposed approach	Highest Accuracy	35	98.00
Proposed approach	Top-10 Features	10	96.83
Proposed approach	Selected Features	23	97.78

Therefore, based on the results, it is reasonable to conclude that our proposed feature selection approach can reduce feature dimensionality effectively. In addition, the framework can also select feature sets flexibly to meet different requirements.

6 Conclusion

In this work, based on the latest phishing dataset, a phishing feature selection approach integrated with Random Forest is proposed, where existing 3 feature importance methods are used to find an effective subset of features. The selected features can reduce the feature dimensionality while maintaining high detection accuracy.

Determining the optimal number of features automatically is a significant task in the selection approach. In our experiment, we define an "Anti-Phishing" Score, which is the weighted average of the four main evaluation metrics based on different requirements. This evaluation score can still be optimized because the feature extraction (preprocessing) time cost is not included. By considering more factors into the "Anti-Phishing" Score, aiming at defining a more convincing evaluation score for deciding the number of features selected.

Further research should also be devoted to the development of practical application like browser extension, which can make it easier and faster for users to identify phishing websites instantly.

In future, researchers may explore a standard set of features which can be worldwide accepted, thus, classification performance can be carried out with less complexity.

References

1. Anti-Phishing Working Group. https://apwg.org/
2. APWG: Phishing Activity Trends Report, 1st Quarter 2021. https://docs.apwg.org/reports/apwg_trends_report_q1_2021.pdf
3. Sahingoz, O.K., Buber, E., Demir, O., Diri, B.: Machine learning based phishing detection from URLs. Expert Syst. Appl. **117**, 345–357 (2019). https://doi.org/10.1016/j.eswa.2018.09.029. ISSN 0957-4174
4. Jain, A.K., Gupta, B.B.: A machine learning based approach for phishing detection using hyperlinks information. J. Ambient. Intell. Humaniz. Comput. **10**(5), 2015–2028 (2018). https://doi.org/10.1007/s12652-018-0798-z
5. Peng, T., Harris, I., Sawa, Y.: Detecting phishing attacks using natural language processing and machine learning. In: 2018 IEEE 12th International Conference on Semantic Computing (ICSC), pp. 300–301 (2018). https://doi.org/10.1109/ICSC.2018.00056
6. Korkmaz, M., Sahingoz, O.K., Diri, B.: Detection of phishing websites by using machine learning-based URL analysis. In: 2020 11th International Conference on Computing, Communication and Networking Technologies (ICCCNT), pp. 1–7 (2020). https://doi.org/10.1109/ICCCNT49239.2020.9225561
7. Wu, C.-Y., Kuo, C.-C., Yang, C.-S.: A phishing detection system based on machine learning. In: 2019 International Conference on Intelligent Computing and its Emerging Applications (ICEA), pp. 28–32 (2019). https://doi.org/10.1109/ICEA.2019.8858325
8. Harinahalli Lokesh, G., BoreGowda, G.: Phishing website detection based on effective machine learning approach. J. Cyber Secur. Technol. **5**(1), 1–14 (2021). https://doi.org/10.1080/23742917.2020.1813396
9. Shahrivari, V., Darabi, M.M., Izadiar, M.: Phishing detection using machine learning techniques (2020). https://arxiv.org/abs/2009.11116
10. Odeh, A., Keshta, I., Abdelfattah, E.: Machine learning techniques for detection of website phishing: a review for promises and challenges. In: 2021 IEEE 11th Annual Computing and Communication Workshop and Conference (CCWC), pp. 0813–0818 (2021). https://doi.org/10.1109/CCWC51732.2021.9375997
11. Chiew, K.L., Tan, C.L., Wong, K., Yong, K.S.C., Tiong, W.K.: A new hybrid ensemble feature selection framework for machine learning-based phishing detection system. Inf. Sci. **484**, 153–166 (2019). https://doi.org/10.1016/j.ins.2019.01.064. ISSN 0020-0255
12. Shabuddin, S., Sani, N.S., Ariffin, A.K., Aliff, M.: Feature selection for phishing website classification. Int. J. Adv. Comput. Sci. Appl. **11**(4), 587–595 (2020)
13. PhishTank. https://phishtank.com/
14. Vrbančič, G., Fister, I., Podgorelec, V.: Datasets for phishing websites detection. Data Brief **33**, 106438 (2020). https://doi.org/10.1016/j.dib.2020.106438. ISSN 2352-3409
15. Vrbančič, G.: Phishing Websites Dataset. Mendeley Data (2020). https://doi.org/10.17632/72ptz43s9v.1
16. Gangavarapu, T., Jaidhar, C.D., Chanduka, B.: Applicability of machine learning in spam and phishing email filtering: review and approaches. Artif. Intell. Rev. **53**(7), 5019–5081 (2020). https://doi.org/10.1007/s10462-020-09814-9
17. Saarela, M., Jauhiainen, S.: Comparison of feature importance measures as explanations for classification models. SN Appl. Sci. **3**, 272 (2021). https://doi.org/10.1007/s42452-021-04148-9
18. Tan, C.L.: Phishing dataset for machine learning: feature evaluation. Mendeley Data (2018). https://doi.org/10.17632/h3cgnj8hft.1

Phishing Email Detection Using Bi-GRU-CNN Model

Mohamed Abdelkarim Remmide(✉), Fatima Boumahdi, and Narhimene Boustia

Laboratoire LRDSI, Faculté Sciences, Université Blida1, B.P 270, Route de Soumaa, Blida, Algeria
abdelkarimremmide@gmail.com, f_boumahdi@esi.dz, nboustia@gmail.com

Abstract. Phishing attacks are the most frequently used method for attackers to obtain sensitive information from victims or infect their networks, as the number of phishing attacks continues to grow rapidly due to their simplicity and low cost of distribution, as well as the appearance of phishing-as-a-service. Thus, phishing email detection is a critical issue that requires immediate attention, where we have focused on resolving the phishing email detection problem using only email bodies. The current study proposed and trained a model using Bi-GRU and two dimensional CNN, in which words are represented using pre-trained GloVe word embeddings. The experimental results show that our model has achieved 98.44% precision, which shows the effectiveness of our model.

1 Introduction

Since the COVID 19 pandemic, a working policy from home has been adopted and in force. As a result, social media and email have emerged as the primary channels of communication between employees. Cybercriminals exploit this channel to gain access to the victim's network or steal sensitive information. This type of attack is referred to as a phishing attack.

Phishing is the most frequently used attack by cybercriminals because, in which it does not require technical knowledge of software or protocol vulnerabilities. Phishing attacks exploit human vulnerabilities through email, social media, SMS, or phone calls using social engineering techniques. The attacker formulates a legitimate-looking email and sends it to the victim with a malicious URL containing a link to download malware or a fake link to a website designed to steal the victim's credentials. The attacker's primary intent is to trick the victim into clicking the link.

During the COVID 19 pandemic, the attacker took advantage of the situation by sending COVID 19-related phishing emails to victims. On April 16, 2020, Google identified 250 million spam emails and 18 million phishing emails daily [1], which resulted in an 11% increase in phishing in 2020, according to the Data Breach Investigations Report (DBIR) [2]. According to the Anti-Phishing Working Group (APWG), the number of attacks decreased in the first quarter

© The Author(s), under exclusive license to Springer Nature Switzerland AG 2022
H. Ragab Hassen and H. Batatia (Eds.): ACS 2021, LNNS 378, pp. 71–77, 2022.
https://doi.org/10.1007/978-3-030-95918-0_8

of 2021, following a peak in January 2021, when 245,771 attacks occurred in a single month [3].

Due to the significance of the phishing detection problem, considerable researches have been conducted to address it. These researches can be classified into phishing URL detection [4], phishing web page detection [5], and phishing email detection [6], where researchers have not focused on phishing email detection problems. The proposed models can be classified as either a machine learning-based approach, which requires feature engineering that necessitates expertise and manual work, or a deep learning-based approach that mitigates the feature engineering problem.

This paper focuses on detecting phishing emails. Thus, we propose a model based on the Zhou et al. [7] Bidirectional Long Short-Term Memory (Bi-LSTM) and Convolutional Neural Network (CNN) model for text classification. However, rather than using Bi-LSTM, we used Bidirectional Gated Recurrent Units (Bi-GRU) because it trains faster. Our work is limited to the email body to ensure that the model can be reused across multiple social media platforms.

This paper is organised as follows: Sect. 2 discusses related works. Section 3 contains our proposed method. Following that, in Sect. 4, we present the dataset and results of the experiment. Finally, in Sect. 5, we conclude the paper and discuss possible future directions.

2 Related Work

In recent years, email usage has shown significant growth, as well as phishing attacks. Thus, numerous anti-phishing techniques have been proposed to mitigate the problem of phishing emails. These solutions are primarily concerned with detecting phishing URLs [4,8] rather than email text. The most straightforward and widely used approach is the blacklist [9,10], but it has some limitations, including an inability to detect new phishing URLs (zero-day attacks), which occur constantly as the attacker is constantly creating or changing these malicious URLs to avoid detection and tracking. This technique required manual identification and continuous updating of the blacklist, both of which require the expertise of a professional.

Following the development of machine learning, researchers attempted to apply these techniques to phishing detection. Numerous proposals were made, including K-Nearest Neighbor (KNN) [11], Naive Bayes (NB) [12], random forest [13], and Support Vector Machine (SVM) [14]. These methods, however, necessitate manual feature engineering, which requires domain expertise and manual work.

After the success of deep learning and the elimination of feature engineering, the researchers were prompted to apply this approach to the problem of phishing detection. Fang et al. [6] propose an improved RCNN model with multilevel representation for the header and body, as well as an attention mechanism. While Nguyen et al. [15] presented a different representation of the email at the word and sentence level and classified it using hierarchical LSTM with supervised attention, Douz et al. [16] attempted to address the issue of a legitimate-looking email containing a malicious URL, but their solution was not implemented. The work of Barathi Ganesh et al. [17] has concentrated on the use of fastText to obtain both the word embedding and the class. Li et al. [18] used a combination of KNN and K-Means to extend the dataset prior to training the LSTM model. In this work we are focusing on detecting phishing email using only the email body.

3 Proposed Approach

In this paper, we approach phishing email detection as a binary classification problem. The email text is being fed into the classifier to determine whether it is phishing or not. As illustrated in Fig. 1, our proposed model consists of the following components: word embedding, Bi-GRU, and CNN.

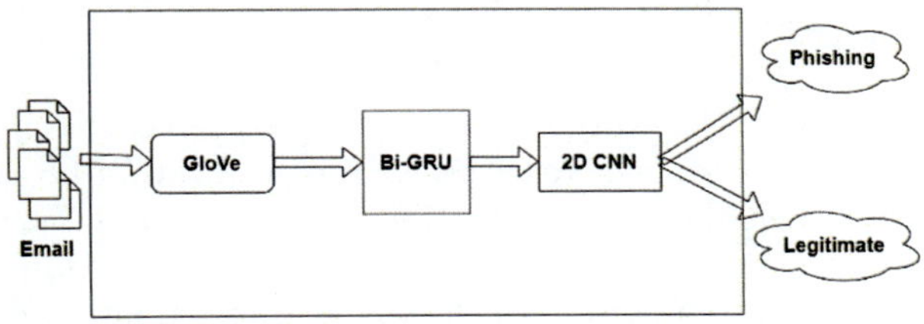

Fig. 1. Our phishing email classification model uses GloVe to generate the vector representation used by Bi-GRU followed by CNN.

3.1 Word Embedding

Due to the use of deep learning, features engineering was avoided. The input data is the raw email body content, which is the portion of the email controlled by people and contains the most information. Deep learning results are optimized by representing the text as a dense vector representation via a word embedding layer. A pre-train model is used to capture the text's syntactic and semantic properties. In which we used in our work GloVe [19] pre-trained word embedding on 27 billion tokens of Twitter tweets.

3.2 Bidirectional Gated Recurrent Units (Bi-GRU)

GRU [20] and LSTM [21] were introduced to mitigate vanishing gradient descent in a simple recurrent neural network (RNN). The distinction between GRU and LSTM is that GRU does not have an output gate. As a result, the GRU structure is more straightforward and requires fewer training parameters. Thus, it theoretically enhances generalization and accelerates training on small data sets while maintaining LSTM like performance. The bidirectional GRU comprises two GRU models, one of which receives input in the forward direction and the other in the reverse direction. So that it can predict the current state using the left and the right state.

3.3 Convolutional Neural Network (CNN)

The Convolutional Neural Network (CNN) is a feedforward neural network primarily used in computer vision [22]. However, it was used in natural language processing (NLP) tasks such as text classification [23]. The CNN network includes convolutional and pooling layers. Commonly in the text classification task, the one-dimensional CNN (one-dimensional convolution and one-dimensional pooling) is used. However, Zhou et al. [7] used a Bi-LSTM followed by a two dimensional CNN (2DCNN) to classify the text, rather than a one dimensional CNN (1DCNN).

3.4 Our Model

Our model is based on the work of Zhou et al. [7], who classified text using Bi-LSTM and a 2DCNN model. However, we replaced the Bi-LSTM model with the Bi-GRU model because GRU has fewer parameters and thus trains faster. The email text is first passed through the GloVe word embedding layer, which generates a dense vector representation. Which is then fed into the Bi-GRU layer to capture future and past information. Following that, the resulted matrix is passed through the two dimensional convolutional layer and the two dimensional max-pooling layer to obtain the complete representation of the input. Following some additional layers such as global maximum pooling and dropout, a dense layer with a sigmoid activation function is used to determine whether the email is phishing.

4 Experimental Evaluation

TensorFlow [24] and Keras are used to implement GPU-accelerated models, Bi-GRU+1DCNN and Bi-GRU+2DCNN. The final chosen hyper-parameters for the two models are presented as follows: word embedding dimension is 100, the hidden units of Bi-GRU are 50, 25 conventional, and 2 max-pooling for Bi-GRU+1DCNN while the other 4. We used a binary cross-entropy loss function and an Adam optimizer with a 0.01 learning rate for Bi-GRU+1DCNN while the other 0.02. The intermediate layer is activated using the ReLu function, whereas the final layer is activated using the sigmoid function. The two dense layers each contain 50 and 30 units, with 0.2 dropouts. The difference between the models is that in Bi-GRU+1DCNN we used a one dimension operation (convolution and pooling) rather than two dimensions. These hyper-parameters can be fine-tuned further to optimize the performance of the model.

4.1 Dataset

All previous research in phishing email detection has been constrained by the lack of a standard and large dataset, whereas the majority of previous research relied on either a private or a combination of open-source datasets. For our experiment, we used a large and unbalanced data from the first Security and Privacy Analytics Anti-Phishing Shared Task (IWSPA-AP 2018) [25], which included two datasets, one with and one without a header. Because we are only interested in the email body, we combine the two datasets by removing the email headers from the first dataset. Next, the duplicate emails are purged. Following that, we preprocessed the dataset by removing punctuation, special characters, and empty spaces and converting the text to lowercase. After that, we split the dataset into an 80% training validation set and a 20% test set using stratified randomization to maintain the same phishing legitimate ratio between sets due to the dataset imbalance.

4.2 Results

The performance of the two models is summarised in Table 1. The accuracy of our model Bi-GRU+2DCNN is 98.47%, 98.84% precision, recall 99.44%, F1-score99.14% and 0.094% FRP. For the other model we have 98.41%, 98.90% precision, 99.32% recall, F1-score 99.12% and 0.056% FRP. The confusion matrix result Table 2 shows that the two models have larger TP and TN with smaller FP and FN, which is the expected value. We observe that the two models have achieved close results, but Bi-GRU+2DCNN results are slightly better than GRU+1DCNN results. However, GRU+2DCNN takes double the time to train the model.

Table 1. Training performance on test set

Model	Accuracy	Precision	Recall	F1-score	FPR
Bi-GRU+CNN	O.9841	0.9890	0.9932	0.9911	0.089
Bi-GRU+2DCNN	0.9847	0.9884	0.9944	0.9914	0.094

Table 2. Confusion matrix

(a) Bi-GRU+1DCNN

		Predicted		
		Positive	Negative	Total
Actual	Positive	1619	11	1630
	Negative	18	183	201
	Total	1637	194	1831

(b)Bi-GRU+2DCNN

		Predicted		
		Positive	Negative	Total
Actual	Positive	1621	9	1630
	Negative	19	182	201
	Total	1640	191	1831

5 Conclusion

This paper presents a phishing detection model that utilizes GloVe word embedding to obtain a dense vector representation of the words. Which is then fed to Bi-GRU with CNN for the classification. After conducting the experiments, we observed that the results for Bi-GRU+1DCNN and Bi-GRU+2DCNN are comparable. The Bi-GRU+2DCNN is superior, but training takes twice as long.

In terms of future work, we intend to continue improving our model for detecting legitimate emails that contain a phishing URL. In addition, to test and improve the system's ability to operate in a real-time social media environment.

References

1. Protecting businesses against cyber threats during COVID-19 and beyond. https://cloud.google.com/blog/products/identity-security/protecting-against-cyber-threats-during-covid-19-and-beyond. Accessed 12 Aug 2021
2. Verizon. 2021 Data Breach Investigations Report 1st Quarter 2021 (2021). https://enterprise.verizon.com/resources/reports/2021-dbir-executive-brief.pdf. Accessed 12 Aug 2021
3. Anti-Phishing Working Group. Phishing Activity Trends Report 1st Quarter 2021 (2021). https://docs.apwg.org/reports/apwg_trends_report_q1_2021.pdf. Accessed 12 Aug 2021
4. Sahoo, D., Liu, C., Hoi, S.C.: Malicious URL detection using machine learning: a survey. arXiv preprint arXiv:1701.07179 (2017)
5. Ding, Y., Luktarhan, N., Li, K., Slamu, W.: A keyword-based combination approach for detecting phishing webpages. Comput. Secur. **84**, 256–275 (2019)
6. Fang, Y., Zhang, C., Huang, C., Liu, L., Yang, Y.: Phishing email detection using improved RCNN model with multilevel vectors and attention mechanism. IEEE Access **7**, 56329–56340 (2019). https://doi.org/10.1109/ACCESS.2019.2913705
7. Zhou, P., Qi, Z., Zheng, S., Xu, J., Bao, H., Xu, B.: Text classification improved by integrating bidirectional LSTM with two-dimensional max pooling. arXiv preprint arXiv:1611.06639 (2016)
8. Tajaddodianfar, F., Stokes, J.W., Gururajan, A.: Texception: a character/word-level deep learning model for phishing URL detection. In: ICASSP 2020–2020 IEEE International Conference on Acoustics, Speech and Signal Processing (ICASSP), pp. 2857–2861. IEEE (2020)
9. Prakash, P., Kumar, M., Kompella, R.R., Gupta, M.: PhishNet: predictive blacklisting to detect phishing attacks. In: 2010 Proceedings IEEE INFOCOM, pp. 1–5. IEEE (2010)
10. Rao, R.S., Pais, A.R.: An enhanced blacklist method to detect phishing websites. In: Shyamasundar, R.K., Singh, V., Vaidya, J. (eds.) ICISS 2017. LNCS, vol. 10717, pp. 323–333. Springer, Cham (2017). https://doi.org/10.1007/978-3-319-72598-7_20
11. Zamir, A., et al.: Phishing web site detection using diverse machine learning algorithms. The Electronic Library ahead-of-print (2020). https://doi.org/10.1108/EL-05-2019-0118
12. Alqahtani, H., Sarker, I.H., Kalim, A., Minhaz Hossain, S.M., Ikhlaq, S., Hossain, S.: Cyber intrusion detection using machine learning classification techniques. In: Chaubey, N., Parikh, S., Amin, K. (eds.) COMS2 2020. CCIS, vol. 1235, pp. 121–131. Springer, Singapore (2020). https://doi.org/10.1007/978-981-15-6648-6_10

13. Akinyelu, A.A., Adewumi, A.O.: Classification of phishing email using random forest machine learning technique. J. Appl. Math. (2014)
14. Figueroa, N., L'huillier, G., Weber, R.: Adversarial classification using signaling games with an application to phishing detection. Data Min. Knowl. Discov. **31**(1), 92–133 (2017). https://doi.org/10.1007/s10618-016-0459-9
15. Nguyen, M., Nguyen, T., Nguyen, T.H.: A deep learning model with hierarchical LSTMs and supervised attention for anti-phishing. In: CEUR Workshop Proceedings, vol. 2124, pp. 29–38 (2018)
16. Douzi, S., Amar, M., El Ouahidi, B.: Advanced phishing filter using autoencoder and denoising autoencoder. In: Proceedings of the International Conference on Big Data and Internet of Thing, pp. 125–129 (2017)
17. Barathi Ganesh, H.B., Vinayakumar, R., Anand Kumar, M., Soman, K.P.: Distributed representation using target classes: bag of tricks for security and privacy analytics. In: CEUR Workshop Proceedings, vol. 2124, pp. 10–15 (2018)
18. Li, Q., Cheng, M., Wang, J., Sun, B.: LSTM based phishing detection for big email data. IEEE Trans. Big Data 1–11 (2020). https://doi.org/10.1109/TBDATA.2020.2978915
19. Pennington, J., Socher, R., Manning, C.D.: Glove: Global vectors for word representation. In: Empirical Methods in Natural Language Processing (EMNLP), pp. 1532–1543 (2014). http://www.aclweb.org/anthology/D14-1162
20. Dey, R., Salem, F.M.: Gate-variants of gated recurrent unit (GRU) neural networks. In: 2017 IEEE 60th International Midwest Symposium on Circuits and Systems (MWSCAS), pp. 1597–1600 (2017). https://doi.org/10.1109/MWSCAS.2017.8053243
21. Sundermeyer, M., Schlüter, R., Ney, H.: LSTM neural networks for language modeling. In: INTERSPEECH (2012)
22. Khan, S., Rahmani, H., Shah, S.A.A., Bennamoun, M.: A guide to convolutional neural networks for computer vision. Synth. Lect. Comput. Vis. **8**(1), 1–207 (2018)
23. Kalchbrenner, N., Grefenstette, E., Blunsom, P.: A convolutional neural network for modelling sentences. arXiv preprint arXiv:1404.2188 (2014)
24. Abadi, M., et al.: Tensorflow: a system for large-scale machine learning (2016)
25. Security and Privacy Analytics Anti-Phishing Shared Task (IWSPA-AP 2018). https://dasavisha.github.io/IWSPA-sharedtask/. Accessed 12 Aug 2021

Securing and Hardening Information Systems

Securing and Hardening Information Systems

Using Physically Unclonable Function for Increasing Security of Internet of Things

Mohammad Taghi Fatehi Khaje[1](✉), Mona Moradi[2], and Kivan Navi[3]

[1] Faculty of Mechanic, Electric and Computer, Science and Research Branch, Islamic Azad University, Tehran, Iran
fatehi@azad.ac.ir

[2] Department of Computer Engineering, Roudehen Branch, Islamic Azad University, Tehran, Iran
mo.moradi@riau.ac.ir

[3] Faculty of Computer Science and Engineering, Shahid Beheshti University, Tehran, Iran
navi@sbu.ac.ir

Abstract. Cryptography is used to secure data and equipment by using a key. The key can also be used for identification and authentication in digital systems. So, the encryption key has a critical role in securing the data and the devices. Therefore, the security of the data and systems depends on the security of the key. A new approach has been proposed, which instead of storing the key in memory, generates it on demand by using an embedded circuit to make the key more secure. This circuit is named Physically Unclonable Function (PUF), which will have a unique output based on process variant and fabrication parameters. The PUF is used as the hardware's fingerprint. This paper will review the research literature and explain why these circuits are suitable solutions for securing the Internet of Things (IoT) and programmable devices with limited resources. This review shows that hardware-based security, being more secure and less costly, is a desirable alternative to traditional software-based methods. We will also present the various types of PUF circuits and explain how this method is resistant to tampering and counterfeiting and how it helps us manage Intellectual property, digital rights, and copyright in the supply chain.

Keywords: Internet of Things · Hardware fingerprint · Physically Unclonable Function · Encryption key

1 Introduction

Nowadays, data is being transferred with incredible speed, making globalization a reality [1, 2]. With the advent of IoT, devices react to surrounding events, exchanging data with each other and with humans. This digital dialogue is happening on the global network of the internet [3]. With the increase of IoT applications, there has been a decrease in the importance of humans' role in gathering and transmitting data. The possibility of accessing devices through the internet means attackers may also gain

© The Author(s), under exclusive license to Springer Nature Switzerland AG 2022
H. Ragab Hassen and H. Batatia (Eds.): ACS 2021, LNNS 378, pp. 81–89, 2022.
https://doi.org/10.1007/978-3-030-95918-0_9

access to information and equipment [4]. In a situation like this, encryption of the data can help secure it, and as a result, protecting the key will be extremely important [5, 6].

Whenever a device sends data, it introduces itself using a key that is used as an ID. The key is usually stored in non-volatile memory, hidden from the user's view, but if an attacker gains access to the device's memory, the security of the key will be at risk [7]. Keys and IDs can be stored in a database and authenticate devices to prevent counterfeiting and illegal cloning [8]. Having said all this, the key's role in the processes of encryption, identification, authentication, validation, and copyright protection should make us attach particular importance to its generation, protection, and storage [9].

Therefore, the generation of and security method used for the key directly affect the security of IoT devices. Making the key more secure will decrease security risks related to these devices [4]. To truly understand the importance of securing IoT devices, one should note that the data and the critical information are being gathered and transferred without direct human supervision or intervention, thus making security assurance of IoT devices using constrained hardware resources a severe challenge [3].

1.1 Hardware-Based Security

In recent years, new hardware-based security solutions for securing devices and encryption keys have been proposed. These methods directly use the output of a circuit as the encryption key, an output which is derived from the hardware by the circuit. Using these approaches can make certain types of attacks harder and more costly [8]. To make data more secure, advanced security methods requiring considerable processing and storage resources are used. It can be a challenge, especially in devices with limited resources. To overcome this challenge, hardware-based security solutions, which do not need complicated processing, thus working with limited resources can be utilized [7].

Because traditional well-known encryption algorithms use private keys for encryption, the whole system's security will depend on the security of the encryption key [10], and protection of the key becomes a severe challenge. In encryption-related computations, considerable resources such as memory, CPU processing, and time are required. In contrast, in IoT devices, we face a severe shortage of such resources for various reasons such as device cost, power usage, and small device size. As a result, hardware-based security solutions can be applied to overcome such resource scarcities.

1.2 PUF as a Key

Considering the limitations on resources in IoT devices, the development of Physically Unclonable Functions (PUFs) is being given special attention. A PUF circuit can extract a unique value as a key from the Integrated Circuit (IC) itself based on the intrinsic properties of the circuit and the variations in the manufacturing process [11].

The key is usually stored in a Non-Volatile Memory (NVM) in digital systems, which means it can be accessible by malicious or non-malicious attacks, and the key may appear and make the encryption ineffective [12]. On the other hand, a PUF circuit extracts the key from the circuit only when needed. As a result, it will make the preservation and generation of the key more secure than when it is stored in an NVM. The PUF automatically generates the response, in real-time, whenever it is requested, and turning off or deactivating the circuit will make it completely inaccessible [6]. In this way, PUF-based methods are immune to physical attacks.

Variations in the manufacturing process are expected in all kinds of fabrication technologies and all types of equipments and physical devices. However, digital devices are being given special attention because of the characteristics of electronic circuit manufacturing and the applications of these circuits [13]. The differences caused by changes in the manufacturing process can be a source of understanding the differences between two products and be used for identifying them [4]. Similar to the way no two people have identical fingerprints or the way the frequency of people's voices can be used to tell them apart, no two pieces of hardware and no two manufactured chips are entirely identical. As illustrated in Fig. 1 PUF is similar to a finger print.

Particular differences of two transistors with the same design made during the manufacturing process can cause differences in the time needed for charging and discharging the capacitors. As a result, no two gates will have the same propagation delay. Such minor differences can be ignored at the level of logical applications, but they can distinguish products [11]. The circuit used to identify differences in manufacturing is called PUF. By activating this circuit, an output is generated, which is used as an identification. The identification is unrepeatable and unclonable because of the complexity of the mathematical model of electronic components [13].

Fig. 1. The similarity between hardware fingerprints and human fingerprints

1.3 Hardware Fingerprint

In situations where the logical and functional structures of the two chips (IC) are the same, there will be minor differences in the manufactured components because of the differences in process variants. The PUF circuit can extract these differences and be used to generate a unique ID, as a digital fingerprint, for each chip [14]. Given that such fingerprints can be generated because of the random differences in the manufacturing process and the intrinsic properties of ICs, even the manufacturers are not capable of producing identical chips. Therefore even the original manufacturers are not capable of cloning the chips [5].

PUF technology, because of its unique characteristics, is an adequate, competitive, and inexpensive IoT security solution. Since it is hardware-based, it has a higher speed of processing and validation and can provide an integrated and secure connection for the internet of things industry [3]. To better understand the difference between software and hardware key, one should pay attention to biometric security systems. Instead of a number or a code as the key, which is accessible by third parties, fingerprints, iris patterns, or other biometric properties can be used.

2 Physical Unclonable Function

2.1 A Brief History of PUF

The term Electronic Physical Random Function was first introduced by Gassend et al. in 2002 at MIT university. They believed that a complicated IC can be viewed as PUF and described this view as a method for independently identifying and authenticating any IC [13]. This concept already existed in optical and magnetic devices production but was not considered an important application [14]. This method of IC identification uses the output of an electronic circuit in place of a private key stored inside the memory or a register. It, therefore, is safe from physical attacks [11].

Because FPGAs are becoming more widely used in digital systems and supply chains and that they are suitable substitutes for ASIC circuits in many applications, manufacturers of digital devices have great hopes for applying PUFs as intellectual property (IP) management tools to prevent illegal cloning of manufactured products and also as means of Digital Rights Management (DRM) [15]. The structure and architecture of PUF circuits are made of transistor circuits and semiconductor technologies. They function based on the propagation delays, the internal functioning of semiconductor components, and the connections between the parts [6].

In the past, different terms were used to refer to PUF circuits. At first, the term "Physical One-way Function" was used, which was not clear enough, and in subsequent researches, they were named Physically Unclonable Functions. However, a simpler term, "Physically Random Functions", was also used. Nevertheless, all this phrases are try to note the fact that the circuits are unclonable, unreproducible and absolutely unique [5]. In some researches, the initial states of the memory cells are introduced as PUF circuits. In [7], dynamic memory cells are used as PUF circuits, and parts of the system's memory are used to generate PUFs.

2.2 PUF Highlights

Examples of researches carried out in the past 20 years about the concept of PUF and the suggested circuits are briefly listed in Table 1. PUF circuits are categorized and classified based on their characteristics.

In [13], the idea is to refer to systems containing parameters affected by variations in the manufacturing process as PUFs. In [6], the applications of PUFs with regards to FPGAs are expanded, and practical solutions are suggested. In [16], the goal is to make the output of PUFs more reliable in various environmental conditions. As a means of IP and copyright protection, new applications for PUFs are proposed in [15]. In [8], the author defends PUF circuits as truly random number generators (TRNGs). In [17], a PUF circuit model based on the CNT-FET technology was simulated. In [1], the idea

Table 1. PUF highlights

Year	Research by	Idea
2002	Gassend et al. [13]	Silicon PUF
2007	Suh and Devadas [6]	RO-PUF on FPGA
2012	Mansouri and Dubrova [16]	Multi-Level Voltage PUF
2015	Barbareschi [15]	PUF for Intellectual Property
2017	Rahman [8]	PUF as TRNG
2017	Moradi et al. [17]	CNT-FET PUF
2018	Tehranipoor [18]	DRAM as PUF
2018	Liu et al. [1]	Reconfigurable RO-PUF
2019	McGrath et al. [14]	PUF Taxonomy
2020	Moradi et al. [19]	Energy-Efficient APUF
2021	Davies and Wang [4]	PUF for Supply Chain Tracking

is to make the PUF circuit more adaptable. In [18], the use of the dynamic memory (DRAM) existing in digital systems is suggested. In [14], many types of research on this subject have been gathered and categorized. The idea in [19] is to show that PUF technologies should reach the level of maturity required to be applied cost-effectively. In [4], the idea of using PUFs in the supply chain is introduced.

2.3 PUF Structure

The structure and architecture of the PUF circuits highlighted in the existing literature can be categorized and classified in various ways [5]. A PUF circuit tries to make explicit the variances caused by the manufacturing process and acquire a unique digital identity by amplifying those variances [1]. For the PUF circuit to be activated, as shown in Fig. 2, an input, termed "Challenge", is applied, and an output, known as the "Response", is received. The pair of the challenge and response is termed "CRP".

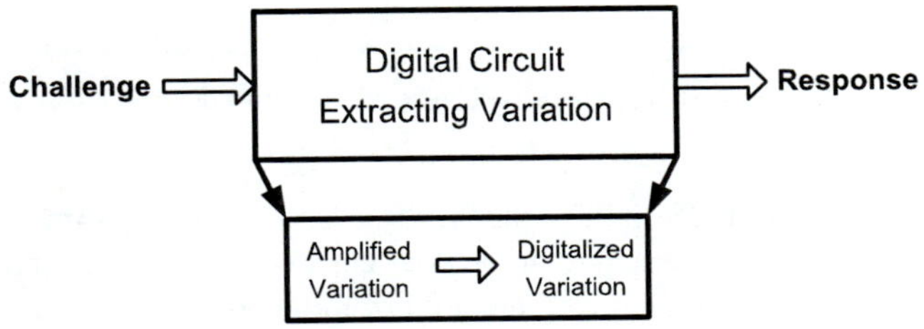

Fig. 2. The general structure of a PUF circuit [1]

Based on past research, PUF electronic circuits can be categorized into three groups [1]; albeit in all such circuits, the randomness and uniqueness result from propagation delays in components, wires, and connection paths:

- Arbiter PUF (A-PUF)
- Ring Oscillator PUF (RO-PUF)
- Memory PUF (RAM-PUF)

The first group of PUF circuits are known as "Arbiter PUFs". Members of this group of circuits are designed based on the variations in propagation delay between different components and their interconnections, and the fact that the route's active changing can create various paths [11]. The output is entirely dependent on the manufacturing process [20]. The general structure of A-PUF circuits is shown in Fig. 3.

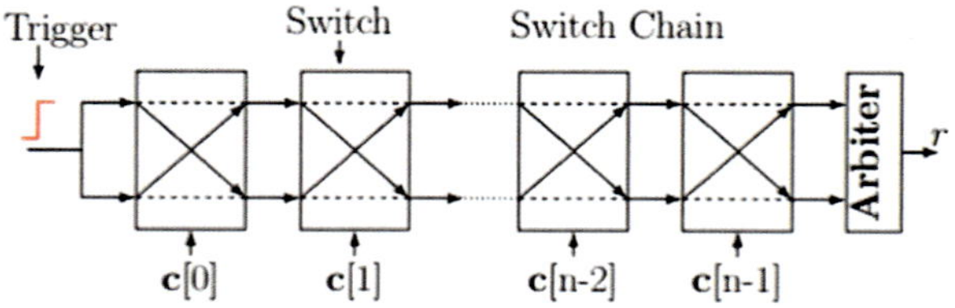

Fig. 3. The functional structure of an A-PUF [20]

Another category of PUF circuits is Ring-Oscillator PUFs. In these circuits, as shown in Fig. 4, an odd number of NOT gates are placed consecutively in a loop-shaped circuit. An oscillatory signal is generated from the changes in gates' outputs counted as the clock [5, 6]. As a result, the generated oscillation, and therefore, the number stored in the counter, will differ for each circuit variant [13, 15].

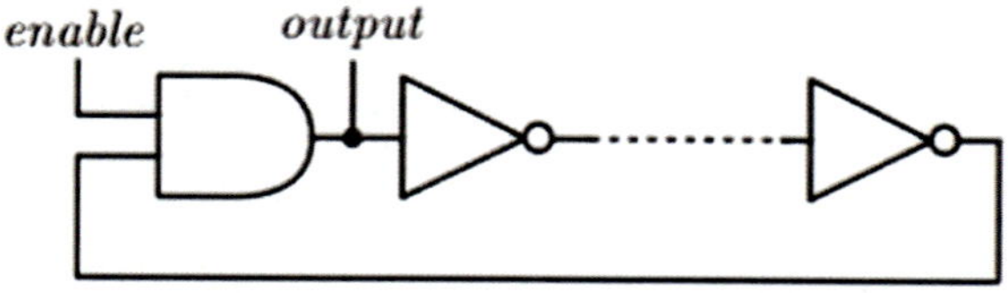

Fig. 4. The general structure of an RO-PUF [15]

The third category of PUF circuits, the "RAM-PUFs", comprises memory elements. In this category, the existing structure of memory cells is used for a different purpose, as a PUF circuit. The general structure of a memory cell is shown in Fig. 5 [18]. The initial value stored in the cell is entirely random, depending on the manufacturing characteristics of the components and the circuit.

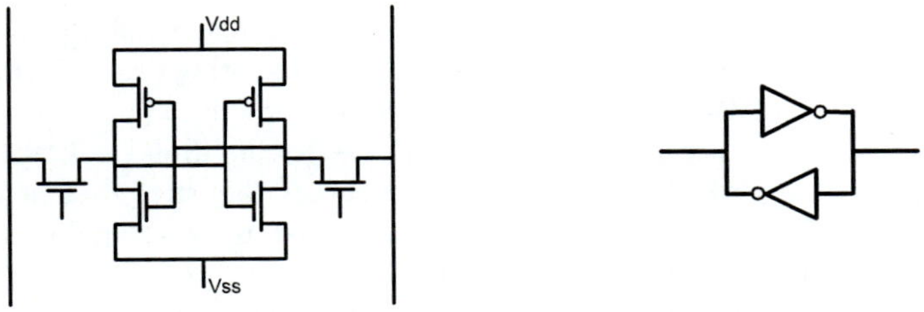

Fig. 5. Electronic and logical memory cell

2.4 PUF's Taxonomy

PUF systems are not limited to electronic circuits. Before the electronic solutions were introduced, they were based on optical and magnetic technologies. One should also note that by combining these solutions, new solutions can be found [13]. In any case, PUF circuits, based on the manufacturing characteristics, can be categorized in various ways.

One approach for such categorizations may be based on whether or not the PUF device is an electronic one. In contrast, in other approaches, whether or not the device's structure can be adjusted, the level of resilience, the methods and concepts utilized, or any other important factor which may separate a group of devices from the others may be the main criteria for grouping the PUFs [14].

No matter how we classify PUF circuits, there will be a list of such circuits to exploit the variations in the manufacturing process properly and create a unique identity for each variant of the designed circuit. Since thay are based on randomness, unpredictability, and unclonability, PUF circuits are evaluated according to these principles. One way of classifying PUFs is by separating them into "Strong PUFs" and "Weak PUFs". If a circuit has a high enough number of such pairs, it will be referred to as an "S-PUF" (Strong PUF) circuit, and if not, a "W-PUF" (Weak PUF) circuit [14, 19].

Another approach groups the circuits into "Intrinsic PUFs" and "Extrinsic PUFs". Intrinsic PUFs distinguish the circuit based on intrinsic characteristics, while extrinsic PUFs exploit external parameters [7]. Another approach for such categorizations is based on the concepts employed in making the PUF circuit and the circuit's areas of application. This approach covers most circuits introduced in the literature [14].

3 The Applications of PUFs

Since hardware circuits are not directly accessible in digital devices, stealing hardware is more difficult than stealing software and data; therefore, hardware circuits are targeted by fewer physical attacks. As a result, hardware-based security solutions such as PUFs have become prominent [3].

In the process of implementing PUF circuits, the intrinsic characteristics of hardware circuits are utilized. Therefore no additional algorithms or resources such as memory or power supplies are needed to provide the intended security. As a result, PUFs are cost-effective security solutions, especially for devices with limited resources [18]. In PUF-based systems, a technique based on the intrinsic characteristics and the propagation delay of integrated circuits (ICs) is defined to identify and authenticate the ICs and based on this technique. As a result of variations in the manufacturing process will cause the ICs to have unique characteristics, the ICs can be mass-produced [11].

Nowadays, digital systems manufactured on a single chip (SoC) and the digital systems implemented using FPGAs have modest resources compared to general-purpose computers [12]. PUFs are utilized to secure them and establish a system for their authentication. Considering the globalization trends, more than ever, manufacturers of digital devices have to face challenges in device fabrication and IP Rights protection [7]. Among the areas in which PUFs have applications are cryptography, identification, authentication, and IP rights protection [9]. The utilization of PUF for digital rights management (DRM), supply chain, and equipment tracking shows its importance in the most novel fields, such as the IoT industry [4]. Six applications of PUF are listed below:

1. PUF as the encryption **key**
2. PUF as the Identification (**ID**)
3. PUF as a means of authentication
4. PUF for proving Intellectual Property (**IP**)
5. PUF for managing Digital Rights Management (**DRM**)
6. PUF for tracking products in the Supply Chain

4 Conclusion

With the widespread use of IoT technologies, hardware products become smaller, with limited resources and more quantities. They are connected to the world outside the internet. Therefore their security has to be guaranteed, and the data transmitted should be encrypted. Moreover, the use of programmable chips to manufacture IoT devices is increasing by the day. Since such devices are clonable, cost-effective solutions are needed to protect intellectual property rights and prevent fabrication. Creation of the certificates and giving access to the equipment is another subject requiring a secure and reliable solution.

Conventional encryption methods based on keys and software algorithms are unsuitable for small devices with limited processing capabilities. As a result, complicated security algorithms are not practical, and less costly solutions are needed. PUF circuits are practical solutions to these requirements, and many such circuits are designed and introduced in the literature.

The considerable amount of research on this subject shows PUF has been highlighted by the researchers for various reasons. Research has shown that PUF circuits are reliable alternatives to software keys used for encryption and device authentication. The unclonability and unpredictability of PUF help IoT devices be more secure.

References:

1. Liu, W., et al.: XOR-based low-cost reconfigurable PUFs for IoT security. ACM Trans. Embed. Comput. Syst. **18**(3), 1–21 (2019). https://doi.org/10.1145/3274666

2. Mukherjee, M., Adhikary, I., Mondal, S., Mondal, A., Pundir, M., Chowdary, V.: A vision of IoT: applications, challenges, and opportunities with dehradun perspective. In: Singh, R., Choudhury, S. (eds.) Proceeding of International Conference on Intelligent Communication, Control and Devices, pp. 553–559. Springer, Singapore (2017). https://doi.org/10.1007/978-981-10-1708-7_63
3. Chatterjee, U., Chakraborty, R., Mukhopadhyay, D.: A PUF-based secure communication protocol for IoT. ACM Trans. Embedd. Comput. Syst. **16**(3), 1–25 (2017)
4. Davies, J., Wang, Y.: Physically unclonable functions (PUFs): a new frontier in supply chain product and asset tracking. IEEE Eng. Manage. Rev. **49**(2), 116–125 (2021)
5. Gassend, B., van Dijk, M., Clarke, D., Devadas, S.: Controlled physical random functions. In: Tuyls, Pim, Skoric, Boris, Kevenaar, Tom (eds.) Security with Noisy Data, pp. 235–253. Springer, London (2007). https://doi.org/10.1007/978-1-84628-984-2_14
6. Suh, G.E., Devadas, S.: Physical unclonable functions for device authentication and secret key generation. In: Proceedings –of Design Automation Conference, pp. 9–14 (2007)
7. Tehranipoor, F.: Design and architecture of hardware-based random function security primitives (2017)
8. Rahman, M.T.: Hardware-based security primitives and their applications to supply chain integrity (2017)
9. Herder, C., Yu, M.D., Koushanfar, F., Devadas, S.: Physical unclonable functions and applications: a tutorial. Proc. IEEE **102**(8), 1126–1141 (2014)
10. Tuyls, P., Škorić, B., Kevenaar, T.: Security with Noisy Data: On Private Biometrics, Secure Key Storage and Anti-Counterfeiting. Springer, London (2007). https://doi.org/10.1007/978-1-84628-984-2
11. Gassend, B., Clarke, D., van Dijk, M., Devadas, S.: Delay-based circuit authentication and applications, p. 294 (2003)
12. Sidhu, S., Mohd, B., Hayajneh, T.: Hardware security in IoT devices with emphasis on hardware trojans. J. Sensor Actuator Netw. **8**(3), 42 (2019)
13. Gassend, B., Clarke, D., van Dijk, M., Devadas, S.: Silicon physical random functions, p. 148 (2002)
14. McGrath, T., Bagci, I., Wang, Z., Roedig, U., Young, R.: A PUF taxonomy. Appl. Phys. Rev. **6**(1), 011303 (2019). https://doi.org/10.1063/1.5079407
15. Barbareschi, M.: Securing embedded digital systems for in-field applications (2015)
16. Mansouri, S.S., Dubrova, E.: Ring oscillator physical unclonable function with multi level supply voltages. In: Proceedings - IEEE International Conference on Computer Design: VLSI in Computers and Processors, pp. 520–521 (2012)
17. Moradi, M., Tao, S., Mirzaee, R.F.: Physical unclonable functions based on carbon nanotube FETs. In: Proceedings of International Symposium on Multiple-Valued Logic, pp. 124–129 (2017)
18. Anagnostopoulos, N.A., Katzenbeisser, S., Chandy, J., Tehranipoor, F.: An overview of dram-based security primitives. Cryptography **2**(2), 1–33 (2018)
19. Moradi, M., Mirzaee, R.F., Tao, S.: CMOS arbiter physical unclonable function with selecting modules (2020)
20. Wisiol, N., Becker, G.T., Margraf, M., Soroceanu, T.A.A., Tobisch, J., Zengin, B.: Breaking the lightweight secure PUF: understanding the relation of input transformations and machine learning resistance. In: Belaïd, S., Güneysu, T. (eds.) CARDIS 2019. LNCS, vol. 11833, pp. 40–54. Springer, Cham (2020). https://doi.org/10.1007/978-3-030-42068-0_3

Check for updates

Multi-face Recognition Systems Based on Deep and Machine Learning Algorithms

Badreddine Alane(✉) and Bouguezel Saad

Department of Electronics, Ferhat ABBAS SETIF1 University, Setif, Algeria
badreddinealane@gmail.com, bouguezel_saad@yahoo.com

Abstract. In this paper, we consider two multi-face recognition systems using deep and machine learning algorithms. Specifically, one is based on the Haar cascade algorithm coupled with the Local Binary Patterns (LBP) classification approach and the other one on the Histogram of Orianted Gradients (HOG) descriptors coupled with the Convolutional Neural Network (CNN) algorithm. We also carry out an exhaustive comparison between the two systems. The simulation results show clearly that both systems have high face detection rates. However, the deep learning-based system outperforms the machine learning-Based system in the face recognition task.

Keywords: Multi-face recognition · Deep learning · Machine learning · Histogram of oriented gradients · Convolutional neural network · Haar cascade · Local binary patterns

1 Introduction

Due to its very important features for identification, the human face has taken a large interest in the biometric field. For instance, face recognition (FR) is extensively used in access control systems, surveillance, identity verification and image database investigations [10]. In general, FR is performed into two successive operations, namely detection and recognition. In the detection step, the face must be localized and extracted from the entire image, whereas the recognition step requires the extraction of some pertinent features of the cropped face to be used as probes to match against know persons features in the gallery [1, 2, 10]. These features can be extracted generally using deep or machine learning algorithm.

There are many applications, such as track gathered attendance and video surveillance, requiring identification of particular persons from a group of people. In this case, the captured entire image contains various faces and hence FR must be applied for each face individually and multi-face recognition becomes crucial.

In this paper, two multi-face recognition systems are considered. One is based on the Haar cascade algorithm, where a cascade function is trained from a number of positive images (images of faces) and negative images (images without faces) [7, 8], and the Local Binary Patterns (LBP) classification approach for boosting statistical local features in the aim of training the system for the FR. The other one is based on the Histogram of Orianted Gradients (HOG) descriptors to construct gradient models for

© The Author(s), under exclusive license to Springer Nature Switzerland AG 2022
H. Ragab Hassen and H. Batatia (Eds.): ACS 2021, LNNS 378, pp. 90–102, 2022.
https://doi.org/10.1007/978-3-030-95918-0_10

detection and the Convolutional Neural Network (CNN) to build a known face database from the right measurements collected by a computer for a high FR accuracy [4]. In addition, we perform a comprehensive comparison between the two systems and show some experimental results.

The reminder of the paper is organized as follows. Section 2 describes the proposed machine learning-based system, whereas the proposed deep learning-system is presented in Sect. 3. The experimental results are shown in Sect. 4 and the conclusion is given in Sect. 5.

2 Deep Learning-Based System

2.1 Histogram of Oriented Gradients

Histogram of oriented gradients descriptors was introduced by [3] for the purpose of human detection. The concept of the descriptor is given as:

- For each pixel from the enter image, we focus on the other pixels that surround it.
- Then, an arrow is drawn towards the pixels, where the image is darker as shown in Fig. 1.
- The designed arrow is called a gradient and it is designed to indicate the darkness direction in the entries of image lighting.
- Gradients are constructed to give an identical image representation for a different image darkness degree for the same person, because pixels will have a different value if luminosity will change even for the same person.
- The process is applied to all the image pixels and the final result of gradient representation is shown in Fig. 2.
- The obtained model is compared to the predefined face HOG model to decide if the area is a face or not.

Fig. 1. The arrow that indicates the side of the darker image zone.

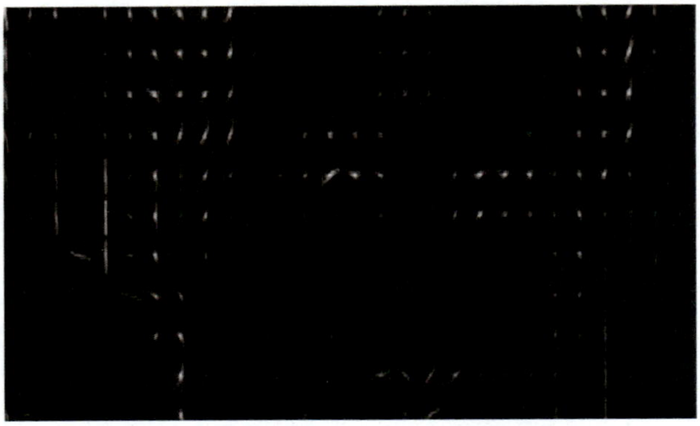

Fig. 2. Face gradient model representation.

2.2 Convolutional Neural Network

The challenge that comes after detecting the face is to find the right measurements to build a known face database. As reported in [4], the successful way for obtaining these measurements to get the best FR performance is to let the computer collecting these measurements by itself. Deep learning does a better job than humans at figuring out which parts of a face are important to measure [4]. So, a deep Convolutional Neural Network (CNN) is trained to collect these needed measurements.

The process of training would treat three face images at the same time:

1. Image of a known person.
2. Another picture of the same known person.
3. A picture of a totally different person.

Then, the algorithm will look at the measurements generated from the images, it adjusts the neural network to make sure that the measurements of the first and second pictures are slightly close and the measurements of the second and third pictures are slightly further apart as described in Fig. 3.

The step is then repeated for each person to be registered in the database, so that the neural network learns to reliably generate 128 measurements for each person [4]. Then, each different picture of a same person would give the same measurements. This process of reducing the image into 128 measurements of each face is called an embedding [4] and the important in this step is not what parts of the face are these 128 numbers measured exactly but the important is that the network generates nearly the same numbers when looking at two different pictures of the same person.

Encoding face images:

This process of training a convolutional neural network to output face embeddings requires a lot of data and computer power. But, once the network has been trained, it can generate measurements for any face, even for the ones it has never seen before. To overcome this disadvantage many networks have been pre-trained, the fine folks at OpenFace did this where several trained networks was published, which can be directly

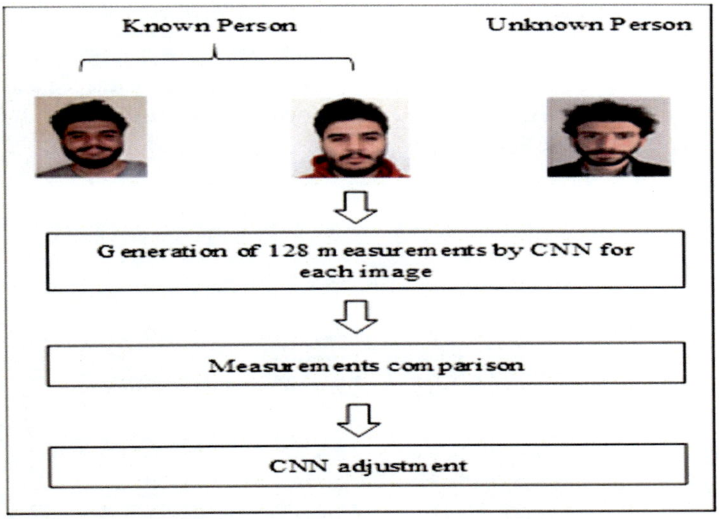

Fig. 3. Convolutional neural network adjustment [5].

used [5]. So, it is only needed to run the face images through the pre-trained network proposed in [5] to get the 128 measurements for each face.

2.3 Proposed CNN-Based system

The proposed machine-learning system is a multi-face detection and recognition system. To implement such a system, preprocessing and training steps must be done before testing any suggested test image, a detailed explanation of the FR system is given below:

1) Face detection

   Each person to be recognized must be trained previously. The training is only made by one picture per person. As presented in Fig. 5, the face is first detected using a HOG descriptor [3] in the training image, then the image is cropped as it holds only the face. This process is repeated with all faces to be recognized.

2) Encoding and registration

   The detected faces must be encoded and registered. Using the pre-trained networks reported in [5], the face is run through the network to get the 128 needed measurements.

   Now, the face is only represented by its 128 generated measurements and it will be registered by its ID and full name. This registration is made by using Structured Query Language (SQL), which helps to access and manipulate databases, repeating this process yield to construct a full dataset to compare the different test images with registered individuals present in this dataset for the recognition task.

   After the three steps (face detection, encoding and registration), the whole system is pre-trained where each individual is represented by his ID and full name in SQL database that will contain all the system needed information for the classification step.

3) Classification

The nearest neighbor classifier [6] with Euclidean distance is used to measure the distance from the test image 128 measurements to the reference image measurements in the gallery. Lower the distance closer the two image measurements. It is defined as (Fig. 4):

$$D = \sqrt{\sum_{m=1}^{128} \left(V_{gallery,m} - V_{test,m}\right)^2} \tag{1}$$

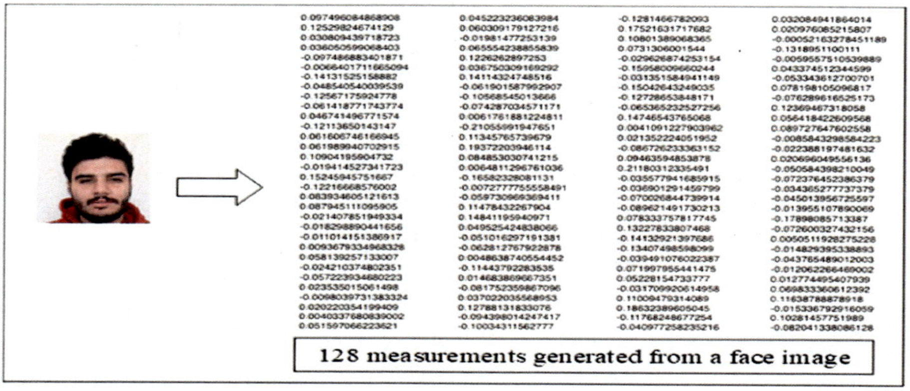

Fig. 4. Generating 128 measurements from a test image.

4) Algorithm of the proposed Face Recognition System

The system was realized using Python language edited on vscode (Visual Studio Code: a code editor redefined and optimized for building and debugging modern web and cloud applications) and the user interface was designed using Qt designer (a Qt tool that provides with a WYSIWYG 'what-you-see-is-what-you-get' user interface to create GUIs for PyQt applications productively and efficiently).

User interface: Qt designer & ptQt.
Registration:

1. Input face image to FR system.
2. Apply the HOG descriptor to get the face zone from the input image and crop the image at its dimensions (Face Detection).
3. Feeding the detected face image to the CNN network to get the vector of the128 measurements (Feature vector) from the face image
4. Saving the extracted feature vector in an SQL database with the corresponding ID.
5. Repeating the steps 1, 2, 3 and 4 for each registration image.

Classification:

1. Input the prob image to FR system.

2. Apply the HOG descriptor for face detection.
3. Multi-Faces will be detected, we'll process next steps for each detected face in the prob image.
4. Extracting feature vector of the128 measurements using the CNN network.
5. Calculating Euclidean distance between the extracted feature vector and all feature vectors present in the SQL database.
6. Selecting the ID of minimum calculated distance and getting the prob face identity (Done for all detected faces in the prob image).

3 Machine Learning-Based System

Emgu CV is a cross-platform image-procession Library [7]. It is closely related to OpenCV because Emgu CV is a.NET wrapper to OpenCV. We can say Emgu CV is OpenCV in.NET. In our system, we have exploited the different pattern recognition algorithms offred by Emgu CV, where we have used the Haar Cascade classifier [4] for the face detection and the LBP descriptor for the training task and both was contained in the Emgu CV library.

3.1 Face Detection Using Haar Cascade

The Emgu CV library offers the use of the Haar cascade algorithm [7, 8] for the process of the face detection. Paul Viola and Michael Jones [8] proposed the first implementation of the face detection technique for the Emgu CV library. The Haar cascade name is due to the use of the Haar functions that calculate the rectangular images containing only the face already existed in the folder.

In the algorithm [8], the Viola-Jones classifier uses a Haar features as an input as showed in Fig. 6. A threshold is applied to the different sum and differences for the rectangular image obtained where the light region is interpreted as 'add region' and the dark one as 'subtract region'. By observing the average people, a rule can be found: the color of the human eye area looks deeper than the cheek area [8]. So, if a Haar feature that can be placed in the adjacent rectangle of the eyes and the cheek area is used then the rectangle can be used for face detection, as in the process in Fig. 6. The next step is to create binary classification nodes of a decision tree, like the Fig. 6. Each of the non-leaf nodes represents a judgment, each path indicates the result of the last judgment, and each leaf represents a kind of output, face or not face. After all the Haar features calculation, the image that passes all the nodes is regarded as a face image [8].

3.2 LBP Approach Classification

The concept of the LBP approach classification is by boosting statistical local features, where the AdaBoost learning algorithm is used to select the Local Binary Patterns based features [9]. The main idea of the LBP pattern recognition is to construct a representation where each pixel is compared to its neighborhood. Let us take the example of a 3 $\times$ 3-pixel bloc, the selection is described in the following steps:

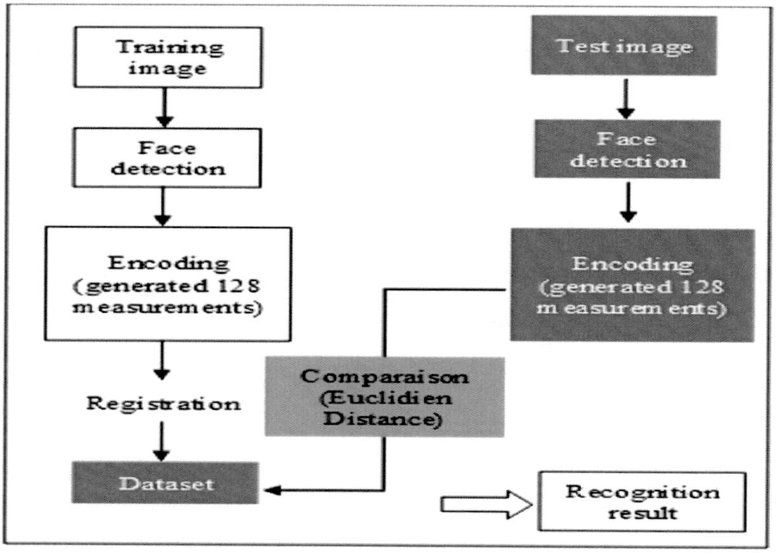

Fig. 5. The diagram of the CNN-based system.

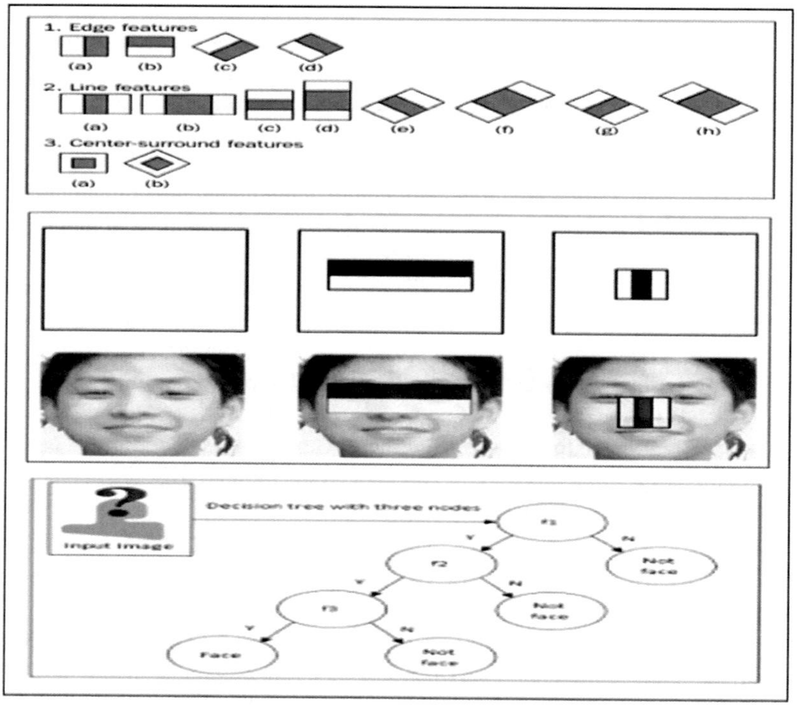

Fig. 6. The Viola-Jones classifier algorithm description.

At first, an intermediate pixel bloc is constructed, we select the centered pixel then a threshold selection is made. If the value of the pixel is higher than the selected pixel, the pixel is set to 0, if the value of the pixel is lower than the selected pixel, the pixel is set to 1. A rotation selection is made to calculate the new pixel value as shown in Fig. 7 to construct an LBP bloc, the bloc contains only the new calculated pixel, see Fig. 7. All the previous steps are then repeated for all the image pixels and the image dataset is created using the LBP pixel images.

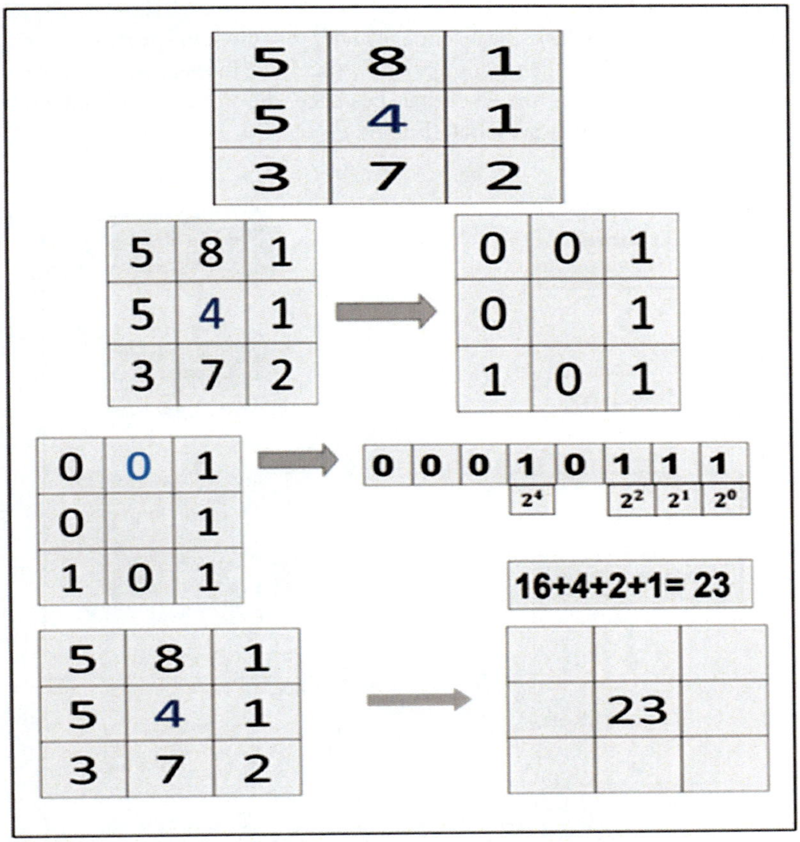

Fig. 7. Elestration of LBP approach classification of 3 × 3-pixel bloc.

3.3 Proposed LBP-Based System

The automatic attendance system is realized as bellow:

1) Registration and image acquisition
   Each student is registered in the database with his personal image and the corresponding information (ID, first/second name). An image of the student is registered

in the database and the other one as a bmp file in the dataset (a gray image for the dataset). The uploaded image may sometimes contain a high luminance so it is converted in a gray image for a better FR performance. This process is the same for the registered dataset image and the test image.

2) Face detection

A Haar cascade classifier is applied to the obtained gray image for detecting the face from it. Then the gray image is reduced to the face frame shape size. This process is the same for the registered dataset image and the test image.

3) Face recognition

For the recognition task, the test image undergoes the same process (gray conversion and face detection) then a comparison with all the faces present in the database is realized by calculating the Euclidian Distance, the minimum distance obtained correspond to the identity searched (Fig. 8).

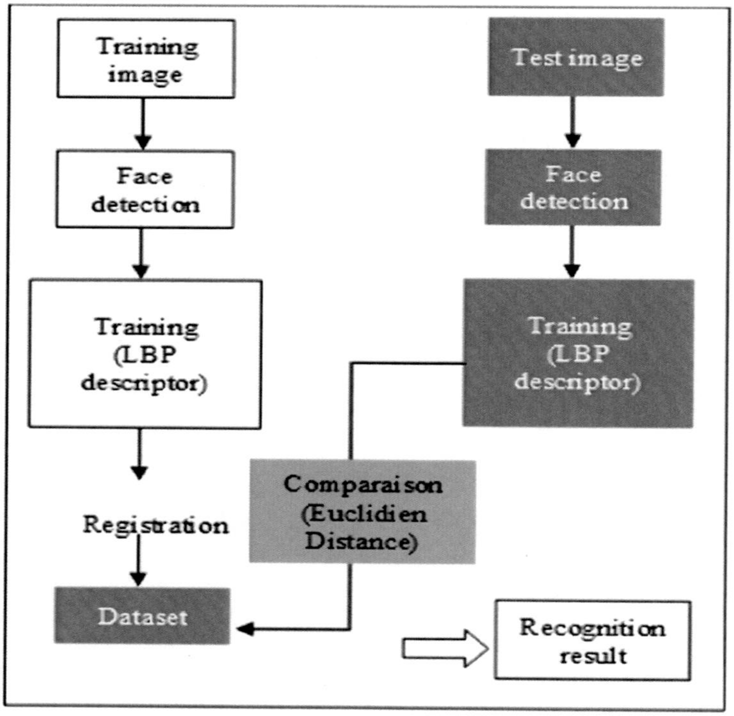

Fig. 8. The LBP-based system diagram.

4) Algorithm of the Proposed face recognition System

The system was realized using C# language edited on Visual Studio (Microsoft IDE used for different types of software development) and the user interface was also designed in.

Registration:

1. Input face image to FR system.
2. Apply the Haar cascade descriptor to get the face zone from the input image and crop the image at its dimensions (Face Detection).
3. Use the LBP pattern recognition algorithm to extract the feature vector from the face image.
4. Saving the extracted feature vector in an SQL database with the corresponding ID.
5. Repeating the steps 1, 2, 3 and 4 for each registration image.

Classification:

1. Input the prob image to FR system.
2. Apply the Haar cascade for face detection.
3. Multi-Faces will be detected, we'll process next steps for each detected face in the prob image.
4. Extracting feature vector using LBP pattern recognition algorithm.
5. Calculating Euclidean distance between the extracted feature vector and all feature vectors present in the SQL database.
6. Selecting the ID of minimum calculated distance and getting the prob face identity (Done for all detected faces in the prob image).

4 Experiment Results

The performances of the two systems are evaluated using a group of voluntary students, the main aim of the experiments is to test the system performance in recognizing the faces present in the test image for different numbers of present students.

The Dataset was not constructed using a familiar database, a group of voluntary university students (12 students) has participated in the construction of our database. Each student was represented by multiple face images then for the test purpose, we consider tested images containing 5, 6, 7, 8, 9, 10 and 12 students.

In the registration step for the CNN-Based system, the individuals were registered by 3, 2 then 1 picture to construct the Dataset and for each registration type (type according to the image registration number) the same process of test was applied, and all types of registration has offered nearly the same results so it was concluded that only one image for registration is sufficient. Then, another characteristic, the quality of the picture, was modified where the registration has been made with a low-quality image, and the result was identical to the one of high-quality image, this advantage has offered a time reducing and low memory requirements.

The image and table of recognition results are shown in Fig. 9 and the recognition rates are given in Table 1 for different registration image numbers and various test images.

Table 1. Recognition rate (%) for CNN-based system.

Registration image number	Individuals present in the test image						
	5	6	7	8	9	10	11
1	100	100	100	100	100	90	100
2	100	100	100	100	100	100	100
3	100	100	100	100	100	100	100

Fig. 9. CNN-based sytem recognition results.

For the LBP-Based system, many pictures of the same person were used for registration (until 10 pictures per person). Although the number of registration images was increased, the system was giving a low accuracy results for the face recognition task unlike face detection. Some image processing techniques (Histogram equalization, gray scale conversion, low pass filter…etc.) were applied on the registration images, but the results are still the same (Table 2).

Table 2. Recognition rate (%) for LBP-based system.

	Individuals present in the test image						
Registration image's number	5	6	7	8	9	10	12
Recognition rate	40	33.33	14.28	25	33.33	20	33.33

Face recognition results are presented below (Figs. 10 and 11):

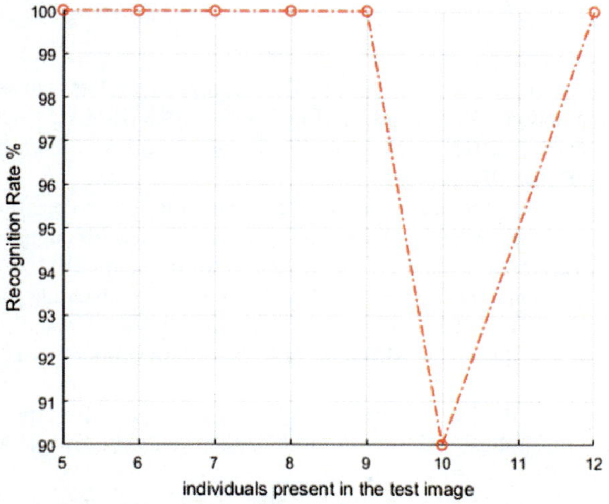

Fig. 10. CNN-based system recognition results.

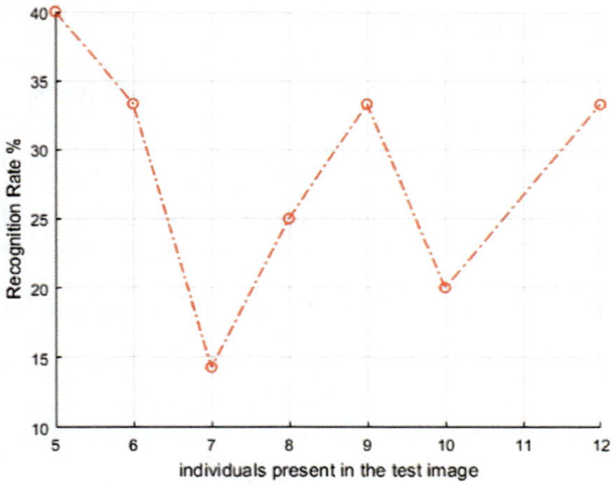

Fig. 11. LBP-based system recognition.

5 Conclusion

In this paper, two multi-face recognition systems have been proposed based on machine and deep learning algorithms. The performances of the two systems, in the detection and recognition, have been discussed and compared. Both systems have offered high face detection accuracy, but the one based on deep learning outperformed that based on machine learning in the face recognition task. This high face recognition performance of the CNN-based system shows how important is to let the machine collects itself the pertinent features of faces.

References

1. Pandian, D., Fernando, X., Baig, Z., Shi, F.: Proceedings of the International Conference on ISMAC in Computational Vision and Bio-Engineering 2018 (ISMAC-CVB) (2019)
2. Wagh, P., Chaudhari, J.: 2015 International Conference on Green Computing and Internet of Things (ICGCIoT). S.l.: IEEE (2015)
3. Dalal, N., Triggs, B.: Histograms of oriented gradients for human detection. In: 2005 IEEE Computer Society Conference on Computer Vision and Pattern Recognition (CVPR 2005), San Diego, CA, USA, vol. 1, pp. 886–893 (2005). https://doi.org/10.1109/CVPR.2005.177
4. Schroff, F., Kalenichenko, D., Philbin, J.: FaceNet: a unified embedding for face recognition and clustering. In: 2015 IEEE Conference on Computer Vision and Pattern Recognition (CVPR), Boston, MA, USA, pp. 815–823, June 2015. https://doi.org/10.1109/CVPR.2015.7298682
5. Amos, B., Ludwiczuk, B., Satyanarayanan, M.: OpenFace: a general-purpose face recognition library with mobile applications. School of Computer Science Carnegie Mellon University Pittsburgh, June 2016
6. Chien, J.-T., Wu, C.-C.: Discriminant waveletfaces and nearest feature classifiers for face recognition. IEEE Trans. Pattern Anal. Mach. Intell **24**(12), 1644–1649 (2002). https://doi.org/10.1109/TPAMI.2002.1114855
7. Shi, S.: Emgu CV Essentials: Develop Your Own Computer Vision Application Using the Power of Emgu CV. Packt Publ, Birmingham (2013)
8. Viola, P., Jones, M.: Rapid object detection using a boosted cascade of simple features. In: Proceedings of the 2001 IEEE Computer Society Conference on Computer Vision and Pattern Recognition, CVPR 2001, Kauai, HI, USA, vol. 1, pp. I-511–I-518 (2001). https://doi.org/10.1109/CVPR.2001.990517
9. Zhang, G., Huang, X., Li, S.Z., Wang, Y., Wu, X.: Boosting local binary pattern (LBP)-based face recognition. In: Li, S.Z., Lai, J., Tan, T., Feng, G., Wang, Y. (eds.) SINOBIOMETRICS 2004. LNCS, vol. 3338, pp. 179–186. Springer, Heidelberg (2004). https://doi.org/10.1007/978-3-540-30548-4_21
10. Winarno, E., Al Amin, I.H., Februariyanti, H., Adi, P.W., Hadikurniawati, W., Anwar, M.T.: Attendance system based on face recognition system using CNN-PCA method and real-time camera. In: 2019 International Seminar on Research of Information Technology and Intelligent Systems (ISRITI), pp. 301–304. Yogyakarta, Indonesia (2019). https://doi.org/10.1109/ISRITI48646.2019.9034596

A Novel Approach Integrating Design Thinking Techniques in Cyber Exercise Development

Melisa Gafic(✉), Simon Tjoa, and Peter Kieseberg

Institute of IT Security Research, St. Pölten University of Applied Sciences, 3100 St. Pölten, Austria
{melisa.gafic,simon.tjoa,peter.kieseberg}@fhstp.ac.at

Abstract. The increasing cyber security compliance requirements (e.g. NIS directive or GDPR in the EU) and the growing dependence on ICT systems have made cyber resilience to a top priority around the globe. Especially during the pandemic, the demand for secure and constantly available systems increased as companies had to change the way they work using home office and remote working capabilities. For satisfying the compelling need for resilient systems, it is vital to ensure that systems work even under adverse events. A key element in this context are cyber exercises. However, planning and conducting an effective cyber exercise is a complex and challenging task. To support this endeavor, in this paper we introduce a novel approach, which integrates design thinking techniques in planning process to improve the exercise development and to tailor exercises to the specific requirements of the organisations. To gain first insights, we evaluated the approach with 50 part-time cyber security students.

1 Introduction

As cyber threats and security incidents continue to take new forms and directions in compromising critical systems, so do security and robustness of these systems continue to be one of the main priorities in the ICT sector, as well as government organisations and public institutions.

In order to respond effectively to the cyber security challenges and to prepare for high-impact cyber risk scenarios, it is necessary to continuously train organisations through exercises. As mentioned in [1], exercises, especially cyber exercises, are an essential component in establishing a resilient society, since they complement to regular preparedness and crisis management exercises, and contribute to continuous improvement.

Cyber exercises typically aim at simulating difficult, complex and realistic situations (e.g. cyber attacks, security incidents) in order to prepare involved organisations for the real case and enable them to react in a more efficient and effective manner. Using different scenarios and through the application of various methodologies, organisations are able to test their ability to protect and recover

© The Author(s), under exclusive license to Springer Nature Switzerland AG 2022
H. Ragab Hassen and H. Batatia (Eds.): ACS 2021, LNNS 378, pp. 103–113, 2022.
https://doi.org/10.1007/978-3-030-95918-0_11

their critical assets in the course of incidents and learn how to mitigate or reduce the impact of such incidents. According to [2], there is no fundamental difference between a simulated event and a real incident in terms of cyber exercises. Cyber exercises can improve reaction time of an organization and reduce the effects of cyber attacks and additionally train participants to handle difficult negative events.

The success of cyber exercises strongly depends on a detailed planning phase considering dependencies, resources and responsibilities. Planning is crucial, since it defines the objectives, guarantees the level of realism of the scenario including various threats, risks and injects. According to ENISA [3], this phase is the most time consuming phase and requires experts' input from various domains. Cyber exercises [4,5] often include non-technical areas of expertise (e.g. legal, media). The inclusion of other disciplines further increases realism as well as the planning complexity.

In order to support exercise planners with new tools, in this paper, we focus on the research question how the cyber exercise planning process can be improved by using design thinking techniques.

The major contribution of this paper lies in introduction of design thinking tools tailored to the cyber exercise planning requirements (e.g. Cyber Exercise Canvas to model the exercise or Personas to describe players and counter-players of the exercise).

The remainder of this paper is structured as follows: In Sect. 2 we briefly survey related guidelines for planning and conducting cyber exercise as well as the design thinking process and toolkits. Section 3 describes our approach for using design thinking methods in cyber exercise planning process. In Sect. 4 we discuss and evaluate our approach. Finally, in Sect. 5 we give a set of conclusions and an outlook on future work.

2 Related Work

In this section, we provide a short overview on existing manuals for supporting conducting cyber exercises and introduce the design thinking approach.

Cyber Exercises

To achieve the desired effect, cyber exercises have to be planned in a way that they are realistic, achievable and tailored to the needs of the organisation(s). For tackling this challenge, the planning phase typically includes multiple stakeholders with various backgrounds. Design thinking methods have already proven to be well suited in other settings when it comes to finding innovative solutions with heterogeneous teams.

To our knowledge this is the first work on structured usage of design thinking methods for the development of cyber exercises. In this section, we firstly present the related work on existing cyber exercise guidelines and standards, which have been an important asset in the development. Secondly, we briefly outline the design thinking process and present existing toolkits. A detailed discussion of the methods used is provided in Sect. 3.

Various standards and guidelines for preparing and conducting cyber exercises already exist. On European level, the European Union Agency for Network and Information Security (ENISA) [3] provides guidance for cyber exercises. The German BSI Standard 100-4 [6] provides even more detailed information on what to consider when organizing and performing an exercise. A publication by RAND Europe [7] deals very intensively with national cyber exercises and their integration into national cyber security strategies.

Guidelines for organizational planning are also available from the MITRE. Their Playbook [8] focuses on practical guidance on cyber exercises. The Homeland Security Exercise and Evaluation Program [9] defines the essential principles to be observed when planning, performing and evaluating exercises within their framework and provides freely accessible templates for all phases. This program aims to be flexible, scaleable and adaptable to address as many stakeholders as possible and can be used for various exercise types and exercise objectives.

The Finnish Manual for Cyber Exercise Organisers [10] describes what cyber exercises are, lists the most important types and describes how regular exercises can be organized. Furthermore, it outlines how exercises support an organisation's preparation for an incident. According to this manual, organisations can combine elements of different types in one exercise. The choice of an appropriate exercise depends on the resources available, the objectives of the exercise and the target group [10].

The Center of Security Studies (ETH Zürich) [2] provides a Cyber Defense Report on cyber exercises. This report examines goals, motivations and benefits of cyber exercises and can be used to derive general metrics.

The Center for Asymmetric Threat Studies (CATS) [1] provides a handbook for planning, running and evaluating information technology and cyber exercises. The handbook highlights ten essential steps of the exercise planning process and also contains practical experiences from previous exercises. In its appendices, example templates and checklists are provided to assist planners.

An interesting suite of cyber security manuals was developed by the National Association of Regulatory Utility Commissioners [11]. This suite contains, among other manuals, a detailed guide to designing and executing a table-top exercise including exercise scenarios and examples.

Design Thinking

Design thinking describes a process supporting the creation of user-centric and innovative products and services. A model commonly used in design thinking is the double diamond model. The process consists of four phases: (i) Discover [12], (ii) Define [13], (iii) Develop [14] and (iv) Deliver [15], alternating between convergent and divergent thinking [16].

Different free toolkits and materials exist, which facilitate the design thinking process. Examples are the MITRE Innovation Toolkit (ITK) [17], consisting of 24 different tools or the repository of design thinking methods and templates by IDEO describing more than 60 different approaches clustered by question or phase [18].

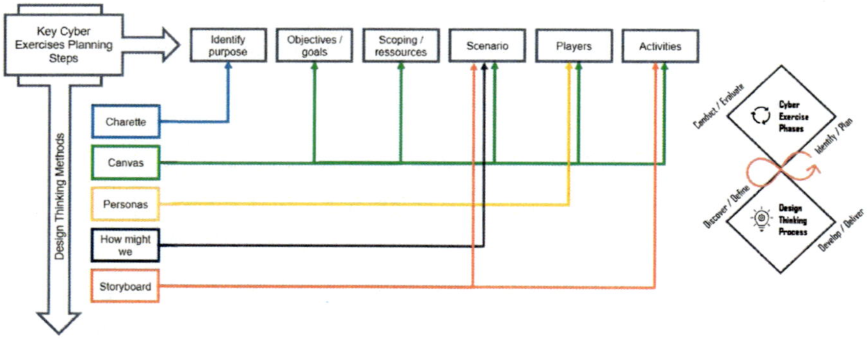

Fig. 1. Integration of design thinking methods into the cyber exercise planning (own representation)

3 Design Thinking for Cyber Exercises

In order to improve the cyber exercise planning processes, we integrated the following design thinking methods into key activities of the ENISA Exercise Identification and Planning Processes [3] (see Fig. 1).

For the identification of the problem, topics and purpose of an exercise, we decided to use the Charette method [19]. In the next step, we propose the usage of an adapted Business Model Canvas [20] to support business case development for the cyber exercise and to determine its essential elements (i.e. goals, objectives, high level scenario, participants, core activities). The visualization of the exercise concept through a Canvas helps to build a common understanding and improves communication during the planning stage.

After determining the cornerstones of the exercise, personas [21] are created to gain a better understanding of the target group (players) on the one side, and the counter-players on the other side. The powerful and simple "How might we"-questions [22] are used to determine how skills can be exercised best. For the refinement of the scenario, the storyboard technique is used [23], which outlines the broad course of events of the scenario and is the foundation for the Master Scenario Event List (MSEL) of the exercise.

Charette Method

Analogue to the design thinking process [24], the first phase of the cyber exercise planning process deals with understanding the exercise context and requirements. Therefore, it is essential to identify the necessity for conducting the exercise and to formulate a typical design question - what is the purpose of that exercise and what are the main objectives?. To exchange and to make existing knowledge about the purpose of conducting cyber exercise visible, we propose to use the Charette method [19].

The Charette method is a synonym for a method of joint, open and public planning [25]. Originally this method was used to discuss and solve urban development problems where citizens were directly involved in decision-making together with project developers [25–27]. However, this method is no longer limited to the field of urban planning and has been applied in many ways as a creative brainstorming technique.

Since Charette does not have a rigid scheme, it can be flexibly adapted to the cyber exercise planning process. Following the Charette procedure, the cyber exercise team is divided into smaller groups. Each group discusses potential security issues and therefore identifies necessity and purpose of the exercise. The discussed points are noted and moved to another group that builds upon these ideas and contributes to the topic with more brainstorming [28]. Once all groups have discussed, they work together to analyze the collected ideas and define steps for the upcoming planning tasks.

The major advantage of this method is that it incorporates a large number of experts with diverse perspectives and ensures contribution of all participants in the planning process. The assumptions and outcomes gathered with the Charette method form the basis for further engagement.

Business Model Canvas

A Business Model Canvas (BMC) is a strategic management tool for visualising and structuring business models and was developed by Osterwalder [29]. To visualize the results gained by the Charette method and to think through vital key components of a cyber exercise, our approach makes use of a variation of the BMC, called Cyber Exercise Canvas.

The introduced Cyber Exercise Canvas consists of nine building blocks (see Fig. 2). Our approach starts with the central element, which are objectives and goals. This key aspect is essential to ensure the effectiveness of every cyber exercise [2,9,30]. Depending on the preference of the exercise planning team, the goals and objectives can be expressed either in a general way (e.g. identification, testing mechanisms and/or procedures...) [2], or in a more detailed way (e.g. grasp the broader national security implications of a wide array of cyber incidents [30], recovering from an incident [10]).

Once the goals are defined, the participants have to be identified. For the specification, an organization-based view (e.g. categorization by sector) [31], or a participant-based view (e.g. technical/operational, management level) [30], can be used.

In the next step, a high level scenario is elaborated. To come up, with an adequate scenario, creativity techniques, such as brainstorming, can be used. A realistic scenario is thereby of vital essence to ensure the success of the exercise [32]. Furthermore, the scenario has to be challenging for the participants to achieve the envisioned learning effects.

After determining the scenario and the participants, the channels allowed for communication during the exercise are determined. The identification of used channels supports the refinement of the scenario and the development of appropriate injects later in the exercise process.

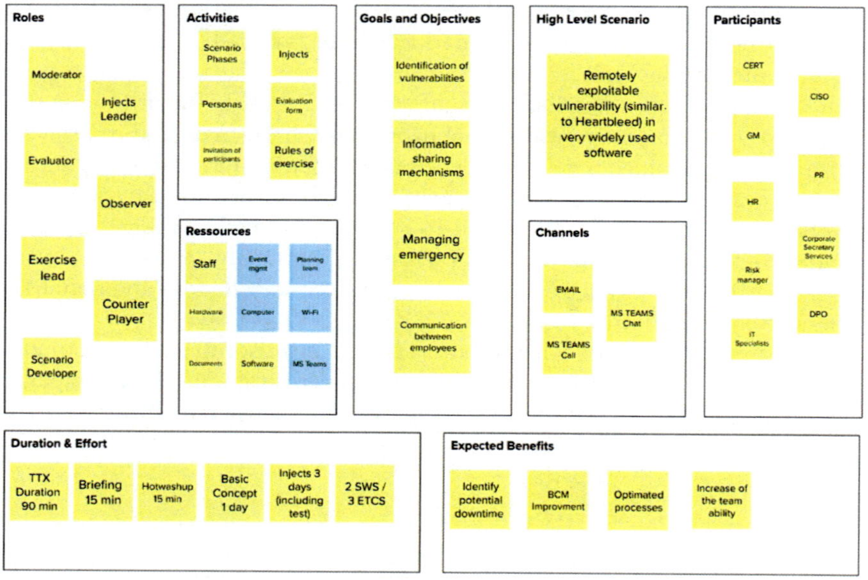

Fig. 2. Cyber Exercise Canvas model (own representation)

The last aspect on the right side of the canvas are the expected benefits. These are important to approach decision makers to gain their commitment and sponsoring as well as to motivate participants to actively take part in the exercise. The next step focuses on the necessary activities, resources and roles, which are necessary to plan, conduct and evaluate the cyber exercise. After these factors are determined, efforts and costs can be roughly estimated and added to the canvas.

Personas

The participants are a central part of the exercise planning process. All planning measures are geared towards the achievement of the exercise objectives by the exercise players and tailored to the skills that should be practised during an exercise. Before the desired participants can be explicitly selected, it is crucial to determine the number of actors and players taken into account, as well as the intended goals, scope and the selected type of exercise [2,3]. Moreover, actors influence performance of an exercise and add realism to the scenario [9].

Exercises in the recent past, such as Cyber Storm [5] and Cyber Europe [33] highlight that usually exercise needs of numerous participants and organisations (from various sectors) have to be satisfied. In order to achieve this objective and to methodically analyse the target group we use the design thinking method called Personas [21,34]. Personas are portraits of fictitious characters created to represent different potential user types that might participate in cyber exercise. The characteristics of these characters are composed of various attributes such

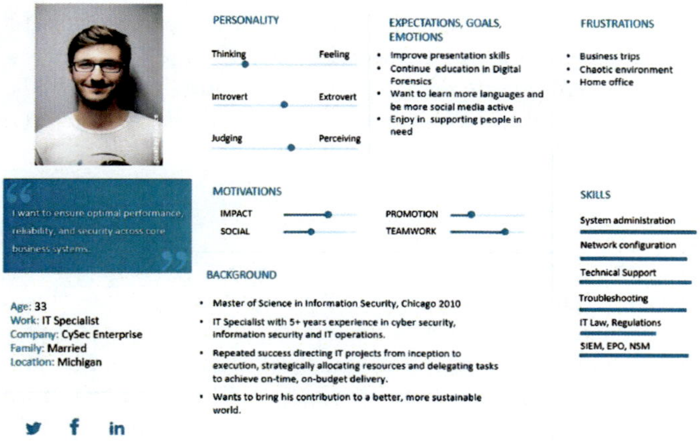

Fig. 3. Persona sample (own representation)

as motivations, frustration, goals and skills and enable us to build empathy with this persona. The more details added, the better we can understand their needs, experiences and behaviours in solving security incidents during cyber exercise. An example of a persona template for cyber exercise participants is depicted in Fig. 3.

- Demographic data (age, work, company, family status, location)
- Background information (education, professional experience)
- Personality
- Expectations, Goals, Emotions, Frustrations
- Motivation and Skills.

"How might we"
Once the Cyber Exercise Canvas is completed and personas are defined, a detailed scenario must be developed to exercise the defined skills and capabilities. To refine the high-level scenario and to find a suitable course of action for the scenario, "How might we" (HMW) questions [22] are a powerful tool.

For making use of the technique, a question is created by "How might we" followed by the goal that should be achieved. An example of such a question could be something like: "How might we test the communication and reaction of the incident response team?".

Storyboard
After collecting ideas how the aims and objectives could be achieved, the basic structure of the exercise scenario has to be defined. To do this, Storyboard is a good technique. Similar to a screenplay, the storyboard contains the most important phases and actions of a scenario.

The goal of this technique is the graphical representation of the scenario to get a common understanding. The process is started by providing empty boxes/cells representing the phases of the scenario. Step by step, each box is filled with a drawing/picture representing the respective phase.

Beside the graphical component of the story board, we enrich the board by also adding essential exercise components, such as a virtual exercise time or textual description of the phase.

4 Evaluation

In order to evaluate our approach, we created standardized questionnaire focusing the user experience. This evaluation form is composed of 12 questions that users answer themselves after using design thinking methods in the cyber exercise planning process. The questions are answered in free text form and a 5 star-rating system, ranging from 5 stars (implying "extremely useful") to 1 star ("not at all useful"). Our evaluation form includes following parts:

1. General part: Overall assessment of conducting the cyber exercise and collaboration in groups.
2. Attractiveness and perspicuity of design thinking methods: Overview of methods and tools used.
3. Efficiency and creativity: How helpful and easy to use were these innovative design thinking solutions?
4. Improvement suggestions: Feedback and suggestions of students to improve our approach.

Within our information security master program, we conducted three cyber exercises including approximately 50 part-time students (working in the security industry) to test our approach. We separated the students into three groups which had to plan and participate in a cyber exercise. The students had a full week to plan, design and run an 1,5 h cyber exercise. In the exercise planning phase, the students applied the introduced design thinking methods and used the Mural tool [35] as an innovative solution for the visual collaboration and as a virtual whiteboard.

During the after-action debriefing of the exercise, respondents received a questionnaire to evaluate efficiency of the applied design thinking methods. First evaluation results are very promising. 69% of respondents have found the Cyber Exercise Canvas extremely and very useful (86% neutral to extremely useful), 55% agree that Personas method was useful for defining target groups and counter players (79% neutral to extremely useful) and overall 83% of respondents have found the Storyboard extremely and very useful to outline scenario phases with corresponding events (100% neutral to extremely useful). The overall feedback was very positive. In the end, students expressed their enthusiasm about creating cyber exercises using creative techniques.

However, further evaluations have to be performed to assess the feasibility in a real-world context with multiple organisations and various target groups (e.g. crisis management team, public relations, legal affairs, IT). We therefore plan to extend our evaluations in academic and professional settings.

5 Conclusions and Future Work

Information and communication technologies are the backbone of our economy. To ensure the resilience of today's organization, it is vital to have proper resilience plans and procedures in place. For assessing the effectiveness of planned responses, the rehearsal of procedures and to test security as well as recovery services, cyber exercises are a central tool.

Designing and planning cyber exercises is a highly complex and difficult task. In order to improve the resulting challenges, a novel approach, which extends traditional exercise design and planning with design thinking components was introduced.

We tested our approach in a first evaluation with approximately 50 students. The students were asked to prepare a cyber exercise within the time frame of a week. A majority of students found the provided design thinking elements useful. We are convinced that the integration of design thinking elements into cyber exercise planning and the design process has huge potential to develop better and more realistic cyber exercises.

The obtained results were still evaluated in a small and superficial setting. In order to eliminate this limitation, in our future work we aim at testing our approach in a real world setting in order to further tailor it to demands of the industry.

References

1. Wilhemson, N., Svensson, T.: Handbook for planning, running and evaluating information technology and cyber security exercises. Swedish National Defence College (2014)
2. Dewar, R.S.: Cybersecurity and cyberdefense exercises. Center for Security Studies - ETH Zürich (2018)
3. ENISA: Good Practice Guide on National Exercises. European Union Agency for Cybersecurity (2009)
4. NATO: Cyber Defence Exercise Locked Shields 2013 - After Action Report (2013)
5. United States Department of Homeland Security: Cyber storm v: after action report (2016)
6. BSI: Notfallmanagement 100:4. Federal Office for Information Security (BSI) (2009)
7. Bellasio, J., Flint, R., et al.: Developing cybersecurity capacity - a proof-of-concept implementation guide. RAND Europe (2018)
8. Kick, J.: Cyber exercise playbook. The MITRE Corporation (2015)
9. FEMA: Homeland Security Exercise and Evaluation Program. Federal Emergency Management Agency (2020)
10. Finnish Transport and N. F. Communications Agency Traficom: Instructions for organising cyber exercises - A manual for cyber exercise organisers. Traficom publication 226/2020 (2020)
11. Costantini, L.P., Raffety, A.: Cybersecurity tabletop exercise guide. National Association of Regulatory Utility Commissioners (2020)

12. Design Council: Design Methods Step 1: Discover, March 2015. https://www.designcouncil.org.uk/news-opinion/design-methods-step-1-discover. Accessed June 2021
13. Design Council: Design Methods Step 2: Define, March 2015. https://www.designcouncil.org.uk/news-opinion/design-methods-step-2-define. Accessed June 2021
14. Design Council: Design Methods Step 3: Develop, March 2015. https://www.designcouncil.org.uk/news-opinion/design-methods-step-3-develop. Accessed June 2021
15. Design Council: Design Methods Step 4: Deliver, March 2021. https://www.designcouncil.org.uk/news-opinion/design-methods-step-4-deliver. Accessed June 2021
16. Design Council: What is the framework for innovation? https://www.designcouncil.org.uk/news-opinion/what-framework-innovation-design-councils-evolved-double-diamond. Accessed May 2021
17. MITRE: MITRE Innovation Toolkit (ITK). https://itk.mitre.org/. Accessed May 2021
18. IDEO: Design Kit - Methods. https://www.designkit.org/methods. Accessed June 2021
19. Farnschläder, L.: Die Charette Methode. https://nativdigital.com/charette-methode/. Accessed May 2021
20. Strategyzer: The Business Model Canvas. https://www.strategyzer.com/canvas/business-model-canvas. Accessed May 2021
21. IBM: Define personas. https://www.ibm.com/garage/method/envision/. Accessed May 2021
22. IDEO: Design Kit - How Might We. https://www.designkit.org/methods/how-might-we. Accessed June 2021
23. IDEO: Design Kit - Storyboard. https://www.designkit.org/methods/storyboard. Accessed June 2021
24. Hasso-Plattner-Institut: Was ist design thinking? https://hpi-academy.de/design-thinking/was-ist-design-thinking.html. Accessed June 2021
25. Bürgergesellschaft. Charette Methodenbeschreibung. https://www.buergergesellschaft.de/mitentscheiden/methoden-verfahren/buergerbeteiligung-in-der-praxis-methoden-und-verfahren-von-a-z/charrette/methodenbeschreibung. Accessed May 2021
26. Mycoted. Charrette. https://www.mycoted.com/Charrette. Accessed May 2021
27. College of Human Ecology in Europe. Die Charette Methode. http://www.coh-europe.de/index.php/de-de/geschichte-2/coh-emmendingen-2/charrette-standortwahl-2/die-charrette-methode-2. Accessed May 2021
28. Elmansy, R.: Brainstorming multiple ideas using charette procedure. https://www.designorate.com/brainstorming-using-charette-procedure/. Accessed June 2021
29. Diehl, A.: Business Model Canvas - Geschäftsmodelle visualisieren, strukturieren und diskutieren. https://digitaleneuordnung.de/blog/business-model-canvas-erklaerung/. Accessed May 2021
30. Ulmanová, M.: How to develop a cyber security table top exercise. National Cyber and Information Security Agency, Technical report (2020)
31. Ogee, A., Gavrila, R., Trimintzios, P., Stavropoulos, V., Zacharis, A.: The 2015 report on national and international cybersecurity exercises - survey, analysis and recommendations. ENISA (European Union Agency for Network and Information Security). Technical report (2015)

32. Department of Homeland Security: Communications-specific tabletop exercise methodology. https://www.hsdl.org/?viewdid=16474. Accessed June 2021
33. E. U. A. for Network and I. S. (ENISA): Cyber europe 2018: After action report (2018)
34. Bjorn, M.: Persona erstellen. https://nativdigital.com/persona-erstellen/. Accessed May 2021
35. Mural tool. https://www.mural.co/. Accessed June 2021

Check for updates

Availability in Openstack: The Bunny that Killed the Cloud

Salih Ismail[1(✉)], Hani Ragab Hassen[2], Mike Just[3], and Hind Zantout[1]

[1] Heriot-Watt University, Dubai, United Arab Emirates
{si8,h.zantout}@hw.ac.uk
[2] Institute of Cybersecurity and Safety, Heriot-Watt University, Dubai, United Arab Emirates
h.ragabhassen@hw.ac.uk
[3] Heriot-Watt University, Edinburgh, UK
m.just@hw.ac.uk

Abstract. The use of cloud is on the rise and is forecast to keep increasing exponentially in the coming years. Openstack is one of the major contributors to the cloud space and there are lot of providers using Openstack for providing cloud solutions. Many organizations run their entire network in the cloud and the availability of the cloud is of paramount importance. Openstack is the popular choice of implementation by many organizations. Openstack is an integration of many projects that make up the platform. There are many advantages to open source modularization and many essential services required to run the infrastructure. One such service is AMQP message broker service and the default one for Openstack is RabbitMQ. Our experimentation shows that it is possible to inject random messages to the queue of RabbitMQ bottling the resource of the main controller. This eventually leads to the entire cloud infrastructure crashing.

1 Introduction

Cloud Computing is an amalgamation of a range of different technologies and about 94% of all enterprises use cloud in one way or another [1]. One of the biggest names of open source cloud Computing infrastructure software is Openstack [2]. Openstack is made up of several projects that work together to provide a seamless experience to the user. Availability is one of the most important features of the cloud, since many critical applications and services of the organizations run in the cloud.

Generally, the most common reason to denying service to a legitimate users is Distributed Denial of Service (DDoS) that involves creating huge amount of traffic or greater speed in the flow of traffic [3]. But to deny service sometimes, all that needs to be done is to take down one essential service that holds the infrastructure together. Such a similar adhesive to the cloud is RabbitMQ. This message broker is based on the Advance Message Queueing Protocol (AMQP) which is one of the essential components in setting up Openstack. It is critical and its setup is one of the major steps performed in the setting of the cloud [4].

© The Author(s), under exclusive license to Springer Nature Switzerland AG 2022
H. Ragab Hassen and H. Batatia (Eds.): ACS 2021, LNNS 378, pp. 114–122, 2022.
https://doi.org/10.1007/978-3-030-95918-0_12

2 Openstack Architecture

2.1 Core Services

Openstack is an open source cloud Operating System that acts as an enabler to an organization to create an IaaS based cloud. Openstack pools all the hardware resources to be provisioned and provides a simple web interface to control the cloud [5]. The ability to deploy Openstack to bare metal, virtual machines or containers makes it desirable due to the variety of options possible as shown in Fig. 1.

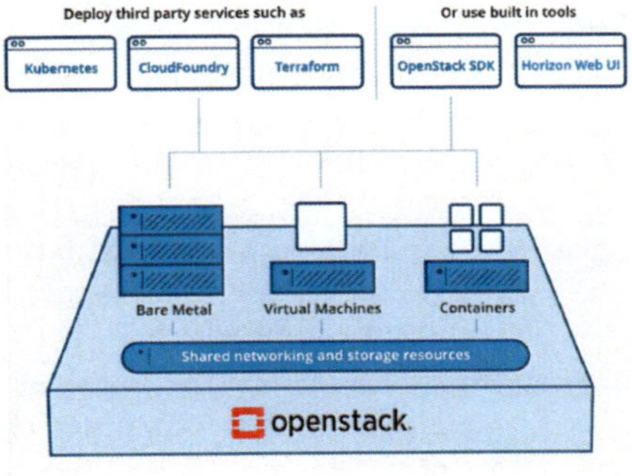

Fig. 1. High level Openstack architecture of Openstack as portrayed in the documentation [5]

Openstack is broken down to smaller sub-projects whereby each of them form the cloud itself. There are four core services that are essential to running a base Openstack environment. Table 1 provides a description of the core services required to run Openstack.

2.1.1 Compute Service

The compute service is referred to as Nova and it implements services and associated libraries to provide massively scalable, on demand, self service access to compute resources. The resources that Nova tries to access include containers, virtual machines and bare metals [6].

2.1.2 Networking Service

Neutron (See Table 1) allows for the creation and attachment of interface devices created and managed by Openstack to other networks. The Physical Network Infrastructure (PNI) and Virtual Network Infrastructure (VNI) are both managed by Neutron. Neutron provides a lot of flexibility to the cloud administrators to create and manage a lot of virtual services in the network like firewall, load-balancers and VPN. The Openstack Network has at least one 'external network',

Table 1. List and details of common Openstack component [7]

Services	Roles	Description	Requirement
Nova	Compute	This service provides scalable, on demand access to compute resources. Some of the responsibilities include spawning, scheduling and decommissioning VMs on demand	Mandatory
Neutron	Networking	Neutron works as SDN project that provides connectivity between the hardware as well as the other Openstack services	Mandatory
Keystone	Identity Service	It essentially works as the security layer of Openstack. The service handles the authentication and authorization of the Openstack services	Mandatory
Glance	Image Service	This service stores the virtual images of the operating systems that helps nova to provision VMs	Mandatory
Cinder	Block Storage	It enables the creation, deletion and management of black storage for the VMs	Optional
Swift	Object Storage	A good solution for storing unstructured data that can grow without a limit	Optional
Horizon	Dashboard	A web based UI that helps in administration of Openstack	Optional

which represents the physical network. On that external network we could have at least one or more 'internal network', that are SDN based virtual networks. This would allow to isolate having smaller clouds within the infrastructure with strong isolation. In terms of security, Openstack allow for the isolation of part of the network, providing for smaller clouds within the infrastructure with strong isolation. Thus, every VM could be a part of one or more Security Groups. Based on the concepts mentioned above, we have two options for creating the virtual network:

1. Networking Option 1: Provider networks: It is the simplest form of networking option in Openstack where it bridges virtual network to physical networks. This option deploys bridging and VLAN segmentation of networks [7].
2. Networking Option 2: Self-service networks: This options routes virtual networks into physical networking using NAT (Network Address Translation). This option allows the customers of the cloud to create virtual interfaces with the involvement of the administrators with help of VXLAN [7].

2.1.3 Identity Service

Keystone service helps to manage credentials in Openstack. Keystone is a single point of integration that uses Openstack's Identity API for client authentication, service discovery and multi-tenant authorization. Openstack maintains a service catalogue that lists all the active services (both mandatory and optional) used by Openstack. These services can have one or many end-points. There are three types of end-points: admin, internal and public. The idea of these different end-points is to provide isolation. For instance, the public end-point can be on a separate network that can be accessed by users through the Internet. This can be further categorization into 'Regions' in Openstack providing scalability. So each Region can have an admin, internal and public end-point providing separation and isolation for better security [8].

2.1.4 Image Service

Glance helps Openstack create, manage and retrieve virtual images. It uses REST API to query metadata of images stored in the database. This query can be used to retrieve the image and launch an instance based on the image. The images can be stored in a simple file system to block based object storage as available in Openstack. There are replication services that runs to ensure that these images are available within the cluster [9].

2.2 Environment Essentials

In order to run the core services listed above, it is important that the following environment components are installed:

- Network Time Protocol: It is important that the services running in controller node and the compute nodes are synchronized. Chrony [10] is recommended as the NTP implementation to be used with Openstack. The suggestion is to use the controller as the reference point for all the other nodes [11].
- Database: Most of the services that run on Openstack use a database for storing information. Generally this service is installed on the controller node. Openstack supports a variety of databases like MYSQL, MariaDB and PostgreSQL [12].
- Caching: Openstack uses memcached to cache the tokens used in the Identity Service. This helps to quicken the process of interaction between the controller and compute nodes [13].
- Key-Value Store: Etcd is used by Openstack for a distributed key locking, storing configuration, keeping track of service live-ness and other scenarios [14].
- Message Broker: RabbitMQ is the deafult message broker used in Openstack. We have a detailed discussion on RabbitMQ in the section below.

Message Broker

The message broken that Openstack uses is RabbitMQ by default for most distributions. RabbitMQ is installed on the controller nodes and sits between the compute nodes to help them to coordinate operations and status information among services. RabbitMQ uses Remote Procedural Calls (RPC) to achieve this. Figure 2 gives a deeper look into the architectural setup of the message broker in Openstack [4].

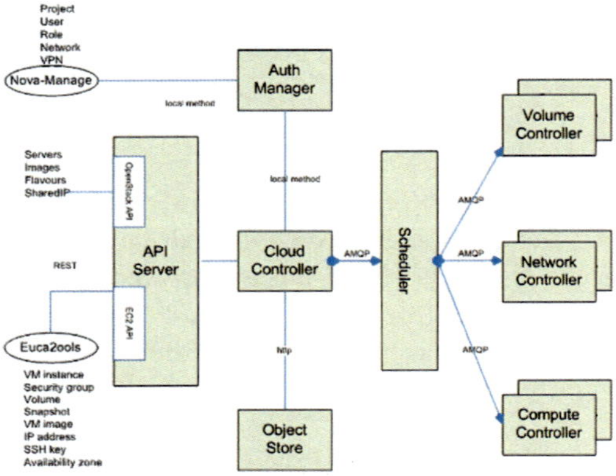

Fig. 2. AMQP and Nova architecture for interaction [4]

Openstack groups and ungroups RPCs into function calls using an adapter. Each nova service creates two queues: one for one which accepts messages with routing keys *NODE-TYPE.NODE-ID* (for example compute.hostname) and another, which accepts messages with routing keys as generic *NODE-TYPE*.

3 Our Experimental Setup

It was interesting to find that not many work was done in relation to the availability study of Openstack setup. Furthermore, papers related to the study of essential components of Openstack was non-existent to the best of our knowledge. Hence, a comparative study was not possible in this particular case. We created an Openstack setup using five machines (Intel Core i5, 8GB RAM) and each machine having two Network Interface Cards. We created two networks: Provider Network and Management Network. We have one machine acting as the controller, two compute nodes, one object storage and one block storage.

There have been exploits in the past that allowed an intruder to get hold of the RabbitMQ authentication details [15]. We argue that the same vulnerability can be extended to be used by an intruder having a simple VM on the cloud. And this would allow the guestVM (currently an attacker), to have full control of the VM which in turn will allow the attacker to take down the entire cloud infrastructure. We created a Proof of Concept in the laboratory with the below assumptions:

- The attacker has been able to successfully attain the authentication credentials.
- The controller where RabbitMQ is installed has a public IP address. A simple IP scan of the range would allow the attacker to gain this information.

We enabled management plugin for RabbitMQ to enable API based access programmatically so that we could alter the variables easily. However, this is not required and this could be done with a simple written script.

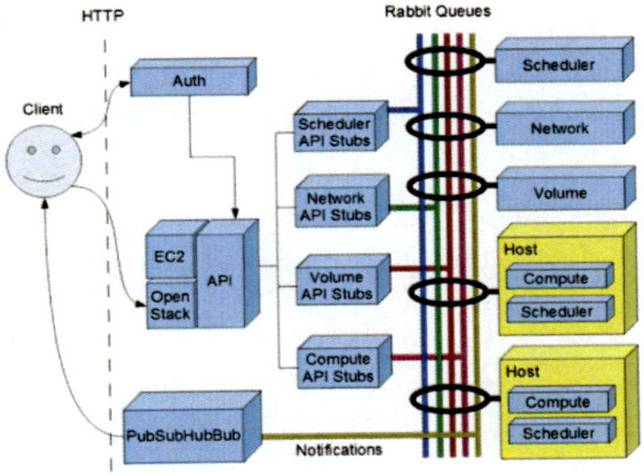

Fig. 3. RabbitMQ in openstack [16]

It is evident from Fig. 3 that the Openstack queues and exchanges should be used for scheduler stubs API, network stubs API, volume stubs API, compute stubs API and notifications. However, We wrote a Java script (the Java API was available for RabbitMQ), where we opened a connection from a VM to the controller. We then created one exchange and multitudes of channels and queues. Through these queues we sent a random message. In order to process this request RAM and CPU resources were being utilized at great level causing the controller to not respond and eventually fail.

4 Results and Discussion

We observed an increase in the number of queues and channels. The message rate rose up to 239 messages/second as shown in the Fig. 4.

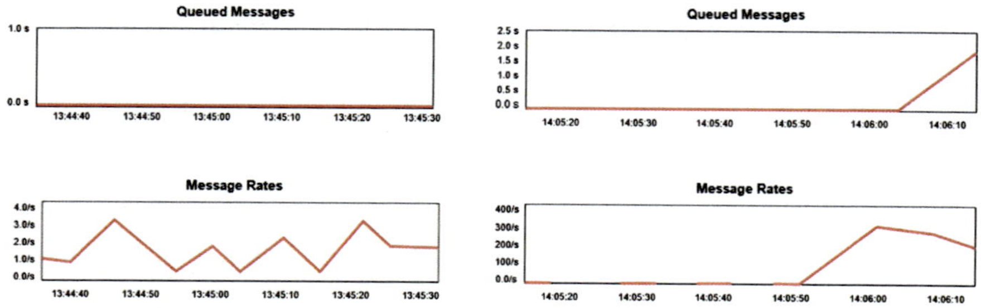

Fig. 4. Comparison of RabbitMQ Message rate and message in queue before (left-hand-side) and during (right hand side) the attack

Table 2 shows the comparison of RabbitMQ variables before and during the attack. As evident there was a single new connection made during the attack and created 4500+ channels and 6100+ channels before eventually the service spiking the RAM and CPU usage of the controller to point of failure.

Table 2. Comparing RabbitMQ variables before and during the attack

	Before the attack	During the attack
Connections	67	68
Channels	67	4618
Exchanges	34	34
Queues	96	6176

We actively monitored the controller's resource at the time to of attack. The results from the attack as shown in Table 3 There was an average increase in 71.3% of the CPU usage. The memory usage simply kept increasing until 99.5% with an increase of 0.1%–0.3% every second. After 99.5% the RAM started using the swap memory of 1 GB which was allowed and this was used up within 23.4 s causing the controller to completely crash.

Table 3. Result of attacking the RabbitMQ service of Openstack on the controller

	Before the attack	During the attack
CPU x 4 (Avg. in %)	12.3	83.6
RAM (Avg. in %)	62.0	99.9
Swap Memory (in Mpbs)	0.0	927.6
Number of Ports Open	25	26

We noticed that the compute nodes were still functional and the VMs resources are utilized from them. But without the controller node and the other environmental essentials which runs on it, the entire range of VMs failed too. This led us to believe that even compartmentalization of the individual services would not work and RabbitMQ failure would act as a single point of failure.

5 Conclusion and Future Work

RabbitMQ which is the default message broker which Openstack Ubuntu distribution uses allowed injection of connections and channels. Ideally, the message broker should limit only the creation of queues and channels only pertaining to Openstack. Everything else should be disallowed. The Simple Authentication Security Layer (SASL) authentication framework does allow for enabling TLS connection. However, if the attacker is able to get hold of the authentication details from the VM as demonstrated in the exploits above, the Proof of Concept would still be effective and disastrous to the cloud. With the correct credentials we could also inject into the existing queues created by Openstack. Putting a limit to these queues and the messages that you pass through them will not be ideal for scalability. This would call for a better method to handle messaging service which essentially connects the different part of the cloud. It would be interesting to further look into those particular activities within the cloud that may be able to create a bottleneck for RabbitMQ and explore the possibility to affect availability.

References

1. Sumina, V.: 26 Cloud Computing Statistics, Facts & Trends for 2021 (2021). https://www.cloudwards.net/cloud-computing-statistics/
2. Openstack: Open Source Cloud Computing Infrastructure - OpenStack (2021). https://www.openstack.org/
3. Elia, I.A., Antunes, N., Laranjeiro, N., Vieira, M.: An Analysis of OpenStack Vulnerabilities (2017). https://bugs.launchpad.nethttps://doi.org/10.1109/EDCC.2017.29
4. OpenStack, OpenStack Docs: AMQP and Nova (2021). https://docs.openstack.org/nova/rocky/reference/rpc.html
5. Openstack: What is OpenStack? https://www.openstack.org/software/
6. Openstack, OpenStack Docs: System architecture (2021). https://docs.openstack.org/nova/rocky/admin/arch.html
7. Openstack, OpenStack Docs: Install Guide Overview. https://docs.openstack.org/install-guide/overview.html
8. OpenStack, OpenStack Docs: Identity service overview (2021). https://docs.openstack.org/newton/install-guide-rdo/common/get-started-identity.html
9. OpenStack, OpenStack Docs: Image service overview (2021). https://docs.openstack.org/glance/rocky/install/get-started.html
10. Curnow, R.: chrony - Introduction (2021). https://chrony.tuxfamily.org/
11. Openstack: Network Time Protocol (NTP) - Installation Guide documentation (2021). https://docs.openstack.org/install-guide/environment-ntp.html

12. Openstack: SQL database - Installation Guide documentation (2021). https://docs.openstack.org/install-guide/environment-sql-database.html
13. Openstack: Memcached - Installation Guide documentation (2021). https://docs.openstack.org/install-guide/environment-memcached.html
14. Openstack: Etcd - Installation Guide documentation (2021). https://docs.openstack.org/install-guide/environment-etcd.html
15. Kirkwood, M.: Bug 1445295 Guestagent config leaks rabbit password: Bugs : OpenStack DBaaS (Trove). https://bugs.launchpad.net/trove/+bug/1445295
16. Openstack: MultiClusterZones - OpenStack (2021). https://wiki.openstack.org/wiki/MultiClusterZones

Distributed and Reliable Leader Election Framework for Wireless Sensor Network (DRLEF)

Nadim Elsakaan(✉) and Kamal Amroun

LIMED – Faculty of Exact Sciences, University of Bejaia, Bejaia, Algeria
{nadim.elsakaan,kamal.amroun}@univ-bejaia.dz

Abstract. The leader election mechanism plays a central role in all technologies requiring automation since the apparition of distributed systems. Indeed, the leader ensures coordination, tasks affectation and charge distribution between the network nodes. Many approaches are proposed in the literature for electing the leader, they commonly present a set of limitations such as the need to go through a spanning tree building stage with the presence of a risk of one point to failure and so on. In this article we present a new distributed algorithm called DRLEF (Distributed and Reliable Leader Election Framework). DRLEF makes use of local information only, it lists the direct neighbors, allows the mapping of the leaders by region and prepares candidates who replace them in case of a failure. The obtained simulation results are very promising.

1 Introduction

Wireless sensor networks (WSN) are a cornerstone technology in the numerical revolution [1], indeed, it is the equivalent of senses for computer and object systems, in a classical layered model, they build the perception layer. A standard hierarchical representation of WSNs decompose them into 02 levels: (i) Gateways: in charge of collecting data and forwarding them to base stations, (ii) Sensors: in charge of collecting data from environment [2,3].

Security services and automation mechanisms are the main challenges facing any technology deployment and achievement, indeed without building a reliable trust model between users and devices it is impossible to realize a wide integration [4,5]. A lot of solutions were proposed to ensure security services in networks context, when focusing on authentication protocols for example, it was found that they are based on a set of assumptions suggesting the presence of a central reliable node called authentication server to authenticate a set of edge nodes [6]. It is worth pointing out that these central nodes are manually chosen among a set of computer or objects with high computation capabilities before deployment and stay inchanged during run time without human intervention. This limitation must be overcome in contemporary networks which aim to increase the speed of recovery from failure by reducing the degree of human intervention. This is where mechanisms like the leader election are required.

© The Author(s), under exclusive license to Springer Nature Switzerland AG 2022
H. Ragab Hassen and H. Batatia (Eds.): ACS 2021, LNNS 378, pp. 123–141, 2022.
https://doi.org/10.1007/978-3-030-95918-0_13

The leader election mechanism plays a central role in automation and auto-management since apparition of distributed systems. It allows designation of a global leader or a set of local leaders in charge of coordinating the work in a group of digital entities, this coordination can take several forms such as; tasks assignment to a set of nodes, distribution of network load, resources allocation for jobs achievement and so on [7]. In this paper, we propose a leader election algorithm for selecting a subset of gateways to play the role of local leaders in charge of authenticating and managing sensors in their range. A lot of works were proposed to take advantage of this mechanism attempting to face some challenges in different application fields, we can cite applications in distributed systems and ad-hoc networks which aim to choose a node as a job manager to attribute tasks to other nodes [9,17,23]. Then it is used generally in IoT context for different purposes [12,19]. Some searchers approach the virtual traffic lights (VTL) problem in context of Vehicular Ad-Hoc Network (VANet) by introducing an election phase in order to choose a car in charge of generating and broadcasting VTL [8,10,18]. Other ones applied this mechanism for choosing a robot leader in order to manage exploration and military tasks without human intervention [13], or for selecting a protocol among population members [22].

The leader election mechanism also finds various use cases in the WSN context; (i) it can be used to organize a number of sensor out of range of any gateway, in order to select one among them which will be in charge of coordinating and finding route to the nearest zone managed by a gateway. Ahcene B et al. [7,11] proposes BROGO an approach which starts by finding a spanning tree starting from the initiator node to allow routing the value of each node to it, then this root node will be in charge of deciding which node is the leader by comparing received values and choosing the minimal one, it is important to mark that the authors assume a flatten network composed only of sensors. (ii) On another hand, due to random deployment, some area can be managed by more than one gateway, they are on charge of authenticating sensors for example, in order to extend lifespan of the WSN we can decide to elect a leader and make others hibernating until the elected one fails. However all these algorithms present the following drawbacks:

- The size of the network impacts the performance of the algorithm, they all start by the search of a spanning tree from the root node which is the initiator of the election process. The duration of this initial phase impacts the global time of the algorithm execution.
- The key role played by the root node leads to one-point-to-failure paradigm. This means that if the root node fails all the protocol execution will fail and the network will take a delay before restarting it.
- They are based on the principle of maximal value which make use of only one information specific to each node such as battery level or computation capacity, this does not considerate the environment of sensors and gateways and is not logical if the algorithm is run few moments after deployment of homogeneous WSN.

In order to avoid these issues, we propose in this paper a new algorithm called DRLEF (Distributed and Reliable Leader Election Framework). It is a new approach addressing the limitations of solutions proposed in the literature, it has the following advantages:

- The use of multiple Gateways by area in order to avoid the one-point-to-failure paradigm.
- The criterion of centrality plays a key role in the designation of a leader.
- An algorithm for calculating disjointed lists in order to identify areas of broadcasts without intersection and whose union is equal to the set of all sensors within reach of the competing gateways. This will allow to send the election messages once and only to each sensor.
- The election result returns a list of candidates to be leader after the elected one fails.
- DERLEF can be generalized for several other use cases because it is applicable for hierarchical systems and it makes use of standard criteria, some adaptations of the parameters and threshold can be sufficient to make it usable in other scenarios.

The rest of this article is structured as follows: Sect. 2 presents the related works that we considered to be the most consistent. Section 3 is devoted to our proposal. Section 4 is dedicated for simulation and performance evaluation. We finish with the Sect. 5 which concludes our work and introduces some interesting perspectives.

2 Related Work

Classically, leader election algorithms are used in distributed systems and collaborative networks in order to allow the designation of a coordinator, in charge of distributing tasks and synchronizing result autonomously. A combination of criteria are used in the election process: computation and storage capacity, networking stability, physical position in the network topology, and so on.

In the IoT context also, leader election finds a lot of beneficial uses, an example is the proposition made, for self-organizing Intersection management, by Christoph Sommer et al. [8]. They tried to propose a leader election based approach to realize Virtual Traffic Lights (VTL), vehicules approaching an intersection exchange messages to organize themselves to avoid collision.

The virtual Traffic Light Algorithm aims to elect a leader for each intersection which is in charge of computing and diffusing a traffic light program to other cars, three main assumptions made this realizable: (i) vehicles embark networking devices as IEEE 802.11p. (ii) each car is equipped with a GPS (Global Positioning System) and supplemented by self-localization methods. (iii) every car maintains a table of neighbors containing IDs (identifiers) of near vehicles. The approach makes use of work of Vasudevan et al. [9] which is an algorithm for dynamic ad-hoc networks and assumes: (i) a specific metric allows ordering nodes: distance to the intersection. (ii) Each node has a unique ID allowing to break ties, the ID can be derived from MAC Address. (iii) Every node keeps track of identifier of current election session.

The VTL algorithm works as follows:

- When a car enters the service area of an intersection, it broadcasts an announcement message containing the distance to the intersection, its own ID. Initially, the initiator considers itself as nearest car to the intersection and exploits a timeout for waiting for replies to the announcement.
- Based on the shortest distance to the intersection, when the timeout is expired, the initiator designates a leader and broadcasts a message to other cars.
- If a car, not actually participating in any election, receives an announcement message, then it engages in current election session and replies with its own distance.
- A car receiving an announcement while participating in an election session joins one which has higher precedence based on the election index or car ID in case of tie.

This algorithm makes use of a specific criteria which is distance, this can be useful in a particular context such as in an ad-hoc network of nodes competing to get access to a critical zone which the roads intersection in this case. In other context like WSN this becomes not usable because of static and heterogeneous nature of nodes. Another reason is the assumption that communications are fully reliable, which cannot be realizable in WSN context. The required architecture for this algorithm involves cars to be equipped with Wireless networking device and localization mechanism such as IEEE 802.11p and GPS. However, it does not make use of crucial features, when in mobility, like speed and acceleration, no mechanisms were proposed to ensure message lose tolerance, when a message is lost the Local Dynamic Map can be corrupted and induces an accident. On the other hand, the criteria to choose the leader is the distance to intersection, vehicle which probably will leave it quickly involving an immediate election restart.

Evaluation mainly concerns intersection crossing time and does not treat of election duration or number of iterations.

Florian Hagenauer et al. [8] proposed an Advanced Leader Election for Virtual Traffic Lights, the algorithm performs next steps when VTL is active at an intersection:

- Vehicles broadcast data about their current position, speed and acceleration.
- If a conflict is detected like an imminent collision, a VTL is generated.
- The closest vehicle to the junction is designated as leader and generates the VTL.
- After crossing the intersection, a new election is started or the leader designates another one before leaving.

Authors evaluated performance of their algorithm based on some important criteria like cars density, travel time and messages loss rate. Once again the criterion used for election is the distance to the junction which cannot be used in other contexts and the election times are not given for different traffic density values.

In [10], the authors proposed a novel algorithm for leader election process in virtual traffic light protocol. Using V2V (Vehicle to vehicle) communications, each vehicle broadcasts its position and speed, if a risk is detected then concerned vehicles follow a protocol, which starts by electing a leader by lane basis the closeness to the intersection. Then the leaders of lanes elect one of them as responsible of traffic lights, it will be in charge of deciding who goes first. Leader can halt or decide to go ahead when its lane shows green light. If it handovers before there are no vehicles waiting at red lights, election process restarts. Authors compared and proves that this VTL approach reduces junction crossing time compared with classical traffic lights from 30 vehicles which is infrequent and does not compared it with any other VTL algorithm.

Some solutions were proposed to agree with WSN context we can cite BROGO (Branch Optima to Global Optimum) as an example, in this approach Bounceur et al. [7] attempt to develop a lightweight leader election algorithm suitable for sensors self-organization.

BROGO works as follows:

- First step consists of running FLF (Flooding for Leaf Finding) algorithm in order to determine a spanning tree, its root and leaves.
- During the second step, each leaf routes a message from itself to the root. This allows routing minimal value of each branch to the reference node which will determine the global minimum.
- In the third and final step, the root node will send a message to the global minimum node informing it that it is the leader.

When looking closer, two main problems appear: (i) what happens if the initiator node, the root of the spanning tree fails, (ii) the FLF algorithm is based on a simple message-acknowledgement, what happens when a node is latent or a message lost.

BROGO considers a flatten WSN with sensor nodes only. It starts by building a spanning tree which allows nodes to communicate with each other. It assumes that messages involved in election procedure must pass by root node. We can easily locate a one-point to failure paradigm in this assumption. Revised BROGO [11] reviews this problem by adding delay, it makes use of Wait-Before-Starting (WBS) procedure. Indeed, if root node does not answer within it, then another node replaces it. In case of failure of leader node, then the election process must start over.

In the wait before start procedure each node identified by x, has to wait a time defined by x * w where nuncio w has to be sufficiently high to allow previous process of informing all nodes. After this waiting time, if no message is received, the root is considered as failed and a second node will start election process. The revision proposes a solution for root failure, using this approach generates important delay and does not solve the problem of lost or latent message (the classical problem expressed as: how to distinguish between latency or loss of a message).

No precision was given in the discussion part, energy consumption was calculated from number of exchanged messages. There are no more details on election phases duration.

Bounceur et al. [12] proposed an algorithm based on a set of local leaders, this approach supposes an arbitrary flatten network, when deployed a set determines nodes with local minima value which are considered as root (local leaders) and start the process of flooding in order to build a spanning tree, when two trees meet then the one with best value continue while the other stops. when the algorithm has run long enough only one spanning tree will remain and its root will be the leader. this algorithm has the advantage of being simplistic but it does not take into account any constraints or any real circumstances Another example where leader election can be used in IoT context, is collaborative networks of robots, Pasquale Pace et al. [13] proposed a Management and Coordination Framework for Aerial-Terrestrial Smart Drone Networks.

A lot of cases require cooperation between aerial and terrestrial robots, in order to coordinate this work team it is necessary to have a leader. The missions are defined outside the group of drones, the leader will be in charge of distributing tasks over collaborative robots. At message receiving from headquarter, nodes broadcast their ability to ensure coordinator role to all other ones within their radio-range. Then, they decide together who will be elected. This leader is not absolute and will be changed according to a specific criterion during time.

The proposition exploits three procedures:

- Look-up: which is executed in specific cases during neighbor discovery phase using multi-cast communication in order to refresh data.
- The leader Election procedure: is used for first leader election and is called again when the leader fails. It is based on multi-cast communication and uses as criterion of selection: maximum remaining charge as maximum ID.
- Mission and task execution: used by the leader to distribute tasks on other nodes in order to complete a specific mission.

This algorithm considers a flatten architecture too (all devices have approximately same capacities), it involves aerial and terrestrial drones in election process, in order to designate a coordinator which will be in charge of distributing tasks to accomplish collaboratively an assigned mission. This objective implies automatically a high mobility model and leading features called to evolve in time, which is a serious challenge we have not to face in WSN context. The total time of election procedure is about 1000 ms when number of nodes reaches 10 which is very high for so few participants and cannot be projected into contexts with thousands of nodes.

Mahendra K.M et al. [14] authors proposed a bio-inspired ant colony approach to realize leader election in context of Cognitive Radio Network (CRN). Following the trend of using ant colonies to achieve leader election in WSN the authors incorporated it into CRNs. They focus on Secondary Users (SU) of which the leader is in charge to listen to communication channels used by Primary Users (PU) and allocate free ones to the SU requiring them.

The performance measures of the algorithm are not conclusive, the comparison was made against old approaches that are not suitable for similar contexts.

3 Distributed and Reliable Leader Election Framework (DRLEF)

3.1 Notations

Table 1. DRLEF variables and messages.

Variable	Role
GWi	Gateway indexed i
LDNNi	Set of sensor nodes in radio range of GWi
NDNNi	Lenght of LDNNi
LDNGi	Set of Gateways in radio range of GWi
NDNGi	Lenght of LDNGi
EIM	Election Initialization Message
DEVij	Deviation of a GWi to a GWj in its radio range
ECG	Election Concerned Gateways List
SEM	Start Election Message
CM	Candidacy Message
ASM	Active Status Message

In this section, we start by listing assumptions which make our algorithm feasible and then we depict step by step the main phases of DRLEF. The Table 1 lists variables used by DRLEF and briefly describes them.

3.2 Assumptions

For our implementation, we assume a classic WSN composed of sensors with limited calculation capacities and gateways which are in charge of data synchronisation, routing and local management.

We assume the following:

- WSN deployment is random.
- As in real WSN, nodes have no high mobility.
- There is a constant ratio between the number of sensors and the number of gateways.
- The deployment area is larger then the radio range of nodes.

An important concept which we exploit in DRLEF algorithm is centrality. A node centrality is a measure of node importance in the network, more so in IoT context where edge nodes play key roles . The most commonly used Centrality node research methods can be classified into the following categories:

1. Between-ness: each node has a score calculated as the fraction of the number of shortest paths passing by a specified node relative to the total number of shortest paths. More the ratio is high more the node is central.
2. Closeness: a node is said to have higher closeness if the sum of its shortest path distance to all other nodes is the smaller one.
3. Degree: takes in consideration the number of direct neighbors of the node.
4. Local Fielder Vector Centrality (LFVC) [15]: measures the network sensibility to a particular node removal.
5. Others: there are a lot of other categories such as Eigenvector centrality and Ego Centrality.

3.3 DRLEF Algorithm

We present here the algorithm DRLEF for leader election, the main objective is to designate a set of Gateways to coordinate the WSN, while other gateways in the same radio-range are hibernating candidates and will wake up in a specific order to replace them if a failure is detected. The full algorithm and its main phases are depicted in the Algorithm 1.

The main phases of DRLEF are:

Algorithm 1: DRLEF Algorithm

```
   Data: Randomly deployed WSN
   Result: Leader per area and ordered successor lists
   // Exploration phase
1  foreach gws_i ∈ list_gws do
2  |  apply a classical flood procedure to detect the direct neighbors of gws_i from
   |    its area;
3  |  if gws_i is alone and there is no another one around then
4  |  |  leave algorithm;
5  |  else
6  |  |  go to initialization phase (line 9);
7  |  end
8  end
   // Initialization phase
9  foreach gws_i ∈ list_gws do
10 |  gws_i creates an EIM (election initialization message) containing its
   |    identifier and a list of its neighbours gateways and sensors;
11 end
12 The gateway with the maximal number of neighbouring gateways is considered
       as CGW (Central) and is in charge of computing the deviation of each
       neighbour gateway; // the deviation is the average of deviations in
       the number of neighbours sensors between the gateway and the
       other gateways in the same area
13 The CGW adds gateways with a deviation greater than a defined threshold to a
     list called ECG (Election Concerned Gateways);
14 The CGW sends a SEM (Start Election Message) containing EGC;
15 if the number of gateways in ECG and in radio range of each other is greater
     or equal to two then
16 |  go to election phase (line 20);
17 else
18 |  leave algorithm;
19 end
   // Election phase
20 foreach gws_i ∈ ECG do
21 |  if gws_i has the minimal score then
22 |  |  the gateways in the same ECG reduces the score of gws_i by the score of
   |  |    the other gateways;
23 |  end
24 |  gws_i compares its score to those of other gateways in the same ECG;
25 |  gws_i sends to the CG (Central Gateway) a CM (Candidacy Message)
   |    containing its final score;
26 end
27 Based on the received scores, the CG computes an ordered list of candidates of
     which the first element is the local leader;
   // Failure tolerance phase
28 repeat
29 |  the elected gws sends an ASM (Active Status Message) to the next gateway
   |    gws_next in the ordered list which will stay in hibernating mode;
30 until there is an interruption for a certain delay;
31 gws_next sends an EM (Elected Message) to the other gateways in the same list
     to inform them that it takes over;
```

1. *Network Exploration*: it is a classical approach to build a local network map, each gateway sends an exploration message and gateways or sensors in radio-range receiving this message, answer with a response containing their network parameters. Then each gateway builds two lists of direct neighbors, one of sensor nodes and the other one of gateways. Note that if a gateway is unique in its own radio-range it is considered as the leader in that network segment as default.

   This phase is summarized by Algorithm 2.

Algorithm 2: Network Exploration

Data: Randomly deployed WSN
Result: Gateways with neighbor lists
foreach $i \in N$ **do**
  //N set of gateway indices
  GW_i sends exploration message to direct neighbors
end
foreach $i \in N$ **do**
  GW_i builds $LDNN_i$, $LDNG_i$; `// List of Direct Neighbor: //sensor nodes (LDNN) and Gateways (LDNG)`
end
if GW_0 *has no gateway as neighbor* **then**
  GW_0 is a local Leader;
  Stop the Algorithm;
end

2. *Initialization*: This phase and following ones occur when two or more gateways are in radio-range of each other. During initialization phase each one of these gateways prepares required lists and information for the election phase. The steps of this phase are:
   - Each gateway prepares an `Election_Initialization_Message` containing its own Identifier with the two lists previously built and then sends this message to all gateways within its radio-range.
   - When all messages received, each gateway checks if it has the maximal number of gateways as direct neighbor, if not, it hibernates and waits for a message.
   - Else the concerned Gateway possessing the maximal number of gateways as direct neighbors is called `Central_Gateway` (CGW), it is important to distinguish this central gateway which is chosen to realize some intermediary calculation, from the elected one which will be designated to lead its zone of the WSN at the end of this algorithm.
   - The CGW calculates deviations of each gateway from each others, the local deviation for a gateway is then calculated as the average value of all individual deviations of this node, the objective is to eliminate GW which has a high average number of sensors different from others. If this value exceeds a threshold, this means that if the concerned gateway hibernates, an important number of sensor nodes will be unreachable, for that reason

we send to this gateways an `Election_Abort_Message` such that they cannot participate to the election. On the other hand, if the deviation is within a reasonable value, we add the concerned gateway to the `Election_Concerned_Gateways` (ECG) list (ordered according to NDNN_i) allowing them to candidate to be the leader.

- The CGW sends a `Start Election Mesage` (SEM) including a key word to start election and the ECG that will be used later by each GW_i to calculate its final list of direct neighbors(sensors), to all gateways in ECG.

Algorithm 3 Describes the operations of this phase.

Algorithm 3: initialization phase

```
Data: Gateways with local knowledge
Result: Local lists of election participants
foreach i ∈ LDNG_i do
    EIM_i={GID, LDNN_i, LDNG_i};                    // GID: Gatway-Identifier
    foreach j ∈ LDNG_i do
        Send(EIM_i,j)
    end
end
// After a waiting delay
if NDNGi ≠ MAX(NDNG_j) then
    Wait for a message reception;
else
    // The concerned Gateway is noted CGW as local central one
    DEV_i=AVGDEV_ij; // The local deviation(DEV_i) of each Gateway is then given as
        the Average of DEV_ij
    if DEV_i > Threshold then
        Send(EAM,i); // EAM: Election Abort Messages
         // GW_i must stay on
    end
    ECG = ECG ⋃{i};
    // ECG: Election Concerned Gateways
    send(SEM,ECG); // SEM: Start Election Message
    }
end
if there are 2 or more gateways in radio-range of each other then
    Goto phase 3 ;
end
```

3. *Election*: Once all lists and candidates are ready, the election phase can start, at the end of which the leaders will be known. It is depicted in Algorithm 4 and steps can be described as follows:
   - On each gateway receiving SEM:
   - Initialize the Final List of Direct Neighbors (LDNF_i, sensor nodes) to the initial one (LDNN_i).
   - For each GW_j from ECG, the GW_i checks if the $\text{NDNN}_j > \text{NDNF}_i$ then it concedes the common sensor nodes to GW_j and $\text{LDNF}_i = \text{LDNF}_i - \text{LDNN}_i$.

- At the end we assume $\text{NDNF}_i = |\text{LDNF}_i|$, each gateway checks if its NDNF_i is greater than Threshold, it sends a candidacy message to the CGW containing its Identifier and NDNF_i.
- When receiving the first CM_i, the CGW waits a predefined laps time for receiving other CM_i and does not accept anymore Candidacies after this delay.
- It creates then an ordered list `Elected List`(EL) according to NDNF_i and attributes rank to each GW_i.
- The GW_i which has the first rank is the elected one (noted EGW) and will stay on.
- CGW sends the EL to all concerned GW_i.
- Other GWi sends a sleep command to sensor nodes in their LDNFi and activate hibernate mode.

Algorithm 4: Election phase

```
Data: Local Lists of Election Participants
Result: Local Leaders
for i ∈ {1, ··· , N} do
    LDNF_i = LDNN_i;
    for j ∈ {1, ··· , Len(ECG)} do
        if NDNF_j > NDNF_i then
            LDNF_i = LDNF_i - LDNN_j;
        end
    end
end
NDNF_i = Len(LDNF_i);
if NDNF_i > Threshold2 then
    CM = {GID, NDNF_i}; // CM: Candidacy Message
    Send(CM, CGW); // CGW: Central GateWay
end
// On CGW when all CM received Estimated by waiting delay from
   first message receiving event
// Create an ordered list according to NDNF_i in descending order
ElectedList = {(GID_1, Rank = 1), ··· , (GID_n, Rank = n)};
Send(ElectedList, GW_i);
if GW_i.rank = 1 then
    ElectedMessage = {GID, Statut ='Elected'};
    Send(ElectedMessage, LDNF_i);
end
HibernateMessage = {GID, Statut ='Not_Elected'};
Send(HibernateMessage, LDNF_i);
```

4. *Failure Tolerance*: This phase introduces an important mechanism which avoids fully restarting the algorithm when the leader fails. It can be decomposed into the following steps:

- Periodically the CGW sends an `Active_Statut_Message` (ASM) to the GW_i with following rank in the Elected List, to inform it that it is always operational.
- If the second gateway does not receive a message within a predefine interval of time from the last ASM, it wakes up and send K periodic `Statut_Check_Messages`, if the CGW does not answer, then the second gateway sends an elected message to gateways and sensor nodes in its radio-range.
- Then this gateway becomes the CGW and the failure tolerance procedure restarts.

Algorithm 5: Failure tolerance phase

```
Data: A local leader
Result: a recovery on failure without re-election
// Current Local Leader is noted EGW as elected gateway
// ASM is an Active Status Message used to inform VGW that EGW is
   still active
// Vice gateway, VGW is one with rank next to EGW
// On EGW:
while True do
    EGW.wait(T) ; // T is a prefixed delay
    Send(ASM, VGW);
end
// On VGW:
while Receiving ASM within T do
    Stay in hibernate mode ;
end
VGW.wakeup();
i = 0;
while (i < k) and (no ASM received) do
    Send(CSM,EGW) ;
    i = i + 1 ;
end
if receiving ASM then
    Restart Failure tolerance phase ;
end
EGW = VGW;
ElectedMessage = {GID, Statut = "Elected"};
Send(ElectedMessage, LDNF_i);
```

4 Simulation

In this section we present the environment of our simulation, the main results of experiments and a brief comparison with existing solutions.

4.1 Environment

Table 2. DRLEF simulation parameters.

Parameter	Value
Radio Range	100 m
Ratio |Sensors|/|gateways|	0.1
Routing protocol	RPL
Threshold	0.2
Threshold2	10

In order to evaluate the performance of our algorithm, we used JBOTSIM framework with parameters described in Table 2 on a machine with the following characteristics:

- Processor: Intel®Core ™i7-4600U CPU @ 2.10GHz 2.70GHz.
- RAM: 8 GO.
- OS: Windows 64 bit, x64-based processor.

JbotSIM is a java library that allows describing, running and evaluating distributed algorithms, it also offers a graphical interface in order to visualize simulation scenarios as they run [16]. Several criteria have been used in the literature to evaluate the performance of election algorithms, some are specific to particular contexts. Pasquele Pace et al. [13] evaluated their management and coordination framework for aerial-terrestrial smart drone networks using average election session duration. We believe this is the best criteria to standardize simulation and allow comparing different approaches regardless of particular scenarios.

We implemented our algorithm and tried several scenarios with variable number of sensors (nodes), while keeping the number of sensors per gateways to 10. We have repeated the experience 50 times for each value in order to get more realistic election duration estimates.

4.2 Results

Our results are summarized in Tables 3, 4 and 5. Each table covers a subset of the entire range of nodes number.

Each row represents a different stage of our algorithm. The first cell of each column gives the number of sensors for its simulation scenario. We kept a constant ratio of 1 gateway for 10 sensors throughout our experiments. Figure 1 to 4 present time evolution according to the number of nodes during the algorithm phases.

Results for exploration phase from Tables 3 to 5 are shown in Fig. 1, the curve follows the form of a second degree polynomial, this observation is also

Table 3. Average election session part 1.

Sensors	100	150	200	250	300	350
Exploration	1,53	3,71	6,36	7,11	7,38	12,96
Initialization	0,6	0,7	0,7	1,2	2,6	2,9
Election	0,3	5,2	11,7	26	45,6	106,9
Total	2,43	9,99	18,76	35,70	55,58	123

Table 4. Average election session part 2.

Sensors	400	450	500	550	600	650
Exploration	17,19	19,18	21,98	27	39,7	50,3
Initialization	1,4	1,3	2,7	1,3	5,4	2,4
Election	157,3	241,2	357,6	460,3	573,2	992,9
Total	157,89	263,86	382,28	492,11	618,3	1036,22

Table 5. Average election session part 3.

Sensors	700	750	800	850	900	950	1000
Exploration	71,8	83,2	104,3	143	169,6	177,3	233,1
Initialization	2,5	4,8	3,1	6,6	4,3	3	15,3
Election	1279	1685,8	1965,1	2650,5	293,4	4265,1	5032,4
Total	1353,3	1696	2072,5	2914,78	2902,6	4619,44	5280,8

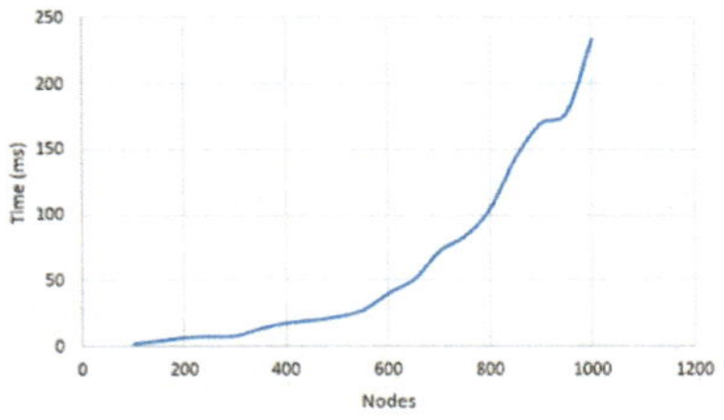

Fig. 1. Exploration phase

valid for the other phases. By being fully distributed our approach remains scalable without a significant over-cost in calculations and therefore in time. As a reminder, the first phase is exploration consisting in messages exchange phase, during which, gateways send requests and wait for responses to build a direct neighbors list of other gateways and sensors.

Figure 3 shows the Election phase results described in Tables 3 to 5, this phase consumes the large share of total time, all calculations, to designate the set of gateways elected as local leaders, are done during this step, based on direct

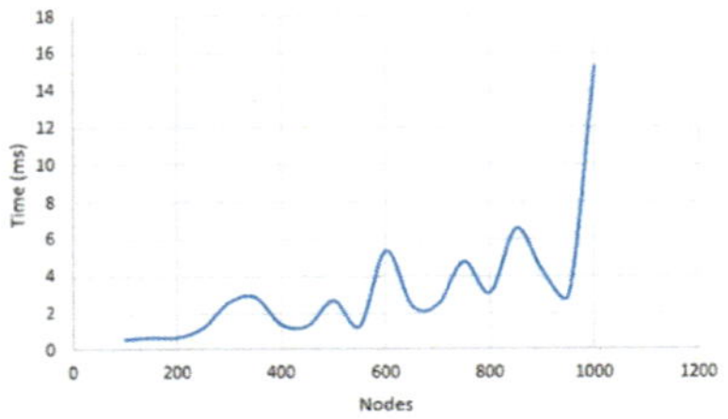

Fig. 2. Initialization phase

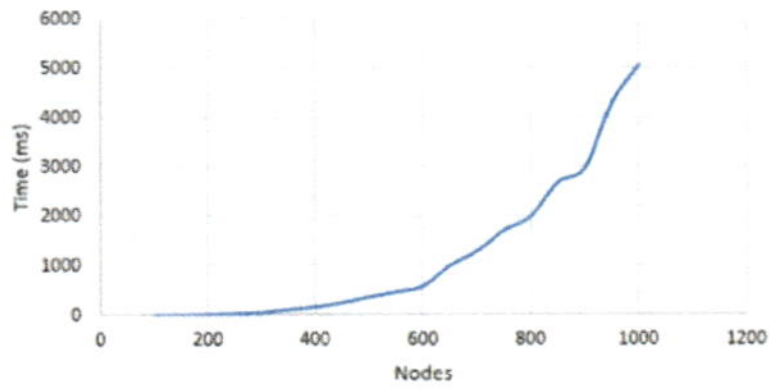

Fig. 3. Election phase

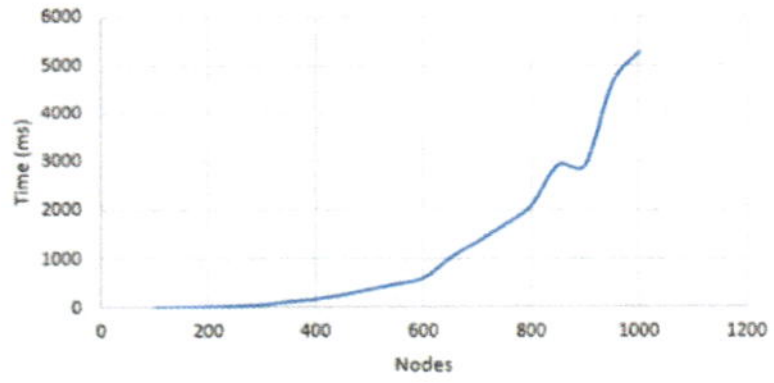

Fig. 4. Total time

neighbors found in phase 1 and using lists prepared and exchanged via messages during the second one.

Figure 4 traces the total time of the protocol, from starting to end of election, globally it remains polynomial, due to the distributed aspect of algorithm execution, multiple instances are running simultaneously in different zone of the WSN topology, which makes its scaling without having a significant impact on the duration of the steps.

Table 6 (traced in Fig. 5) presents a comparison, based on total duration, between DRLEF and three of other leader election algorithms summarized in related Works. The criterion of time is the most indicative in our opinion, it contains: (i) network map building, (ii) calculation time which is in direct correlation with the algorithm complexity, and (iii) messages exchange duration which depends on number of messages used. For all these reasons, we consider that time is the central criterion to evaluate and compare algorithms because it allows estimation of all other criteria. According to the comparison, we can easily notice that our algorithm is less time consuming, and therefore lower in number of messages and requires less computation operations.

Also, we can observe that BROGO [7] gives best performance from thousand nodes, this is because it considers a flatten WSN composed of sensors only (without gateways) which makes its complexity near to the spanning tree algorithms one. it's worth noting that this is not suitable for real situations and cannot be adapted for different situations and scenarios.

Table 6. Algorithms comparison

Sensors	200	400	600	800	1000
DRLEF	68,25	153,33	911,83	2054,14	4531,4
MCFATD [13]	1890	3690	5490	7290	9090
ICNP [9]	2000	5000	7500	12000	14000
BROGO [11]	3400	3500	3600	3700	3800

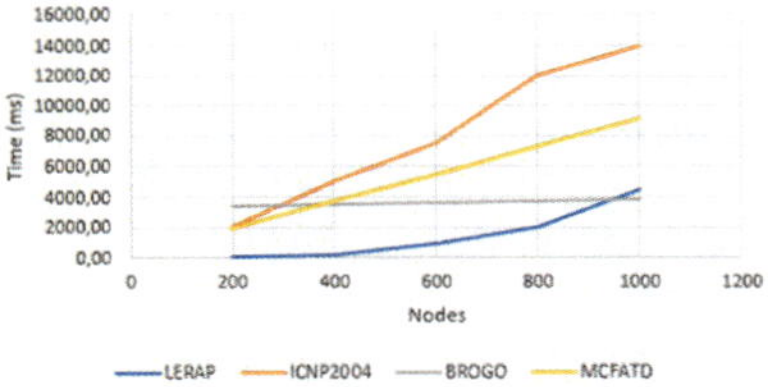

Fig. 5. Algorithms comparison

5 Conclusion and Future Work

This paper presented the design of DRLEF, a lightweight leader election protocol which can be used to designate authentication server among a collaborative group of Gateways. DRLEF makes use of logical criterion, centrality which corresponds to number of direct competetive neighbours, this can be used in different context.

Effectiveness has been proved by simulation which shows that DRLEF ensures scalability with reasonable processing overhead. This is due to the distribution manner in which we have distributed election procedure, instead of building spanning trees and other structures to cover all the WSN, we make elections locally in each WSN zone. In addition we have proposed a failure-tolerance procedure in order to restart election when a leader fails.

As perspective we inspire to add a procedure to manage mobility, indeed, our algorithm supports disconnecting nodes or failing ones, but can not ensure election in extreme mobility conditions.

References

1. Noura, M., Atiquzzaman, M., Gaedke, M.: Interoperability in internet of things: taxonomies and open challenges. Mob. Netw. Appl. **24**(3), 796–809 (2018). https://doi.org/10.1007/s11036-018-1089-9
2. Hasan Ali, K., Munam Ali, S., Sangeen, K., Ihsan, A., Muhammad, I.: Perception layer security in Internet of Things. Future Gener. Comput. Syst. **100**, 144–164 (2019)
3. Jeretta Horn, N., Alex, K., Joanna, P.: The Internet of Things: review and theoretical framework. Expert Syst. Appl. **133**, 97–108 (2019)
4. Aakansha, T., Gupta, B.B.: Security, privacy and trust of different layers in Internet-of-Things (IoTs) framework. Future Gener. Comput. Syst. **108**, 909–920 (2018)
5. Mardiana, M.N.B., Haslina, W.H.: Current research on Internet of Things (IoT) security: a survey. Comput. Netw. **148**, 283–294 (2018)
6. Mohammed, E., Ahmad, F., Maroun, C., Ahmed, S.: A survey of Internet of Things (IoT) authentication schemes. MDPI Sens. **19**(5), 1141 (2019)
7. Bounceur, A., Bezoui, M., Euler, R., Kadjouh, N., Lalem, F.: BROGO: a new low energy consumption algorithm for leader election in WSNs. In: 10th International Conference on Developments in eSystems Engineering (DeSE), pp. 218–223. IEEE, June 2017
8. Sommer, C., Hagenauer, F., Dressler, F.: A networking perspective on self-organizing intersection management. In: 2014 IEEE World Forum on Internet of Things (WF-IoT), pp. 230-234. IEEE, March 2014
9. Vasudevan, S., Kurose, J., Towsley, D.: Design and analysis of a leader election algorithm for mobile ad hoc networks. In: Proceedings of the 12th IEEE International Conference on Network Protocols, ICNP 2004, pp. 350-360. IEEE, October 2004
10. Choudhary, P., Dwivedi, R.K., Singh, U.: Novel algorithm for leader election process in virtual traffic light protocol. Int. J. Inf. Technol. **12**(1), 113–117 (2019). https://doi.org/10.1007/s41870-019-00305-x
11. Bounceur, A., Bezoui, M., Euler, R., Lalem, F., and Lounis, M. A revised BROGO algorithm for leader election in wireless sensor and IoT networks. In: 2017 IEEE Sensors, pp. 1–3. IEEE, October 2017
12. Bounceur, A., Bezoui, M., Lounis, M., Euler, R., Ciprian, T. : A new dominating tree routing algorithm for efficient leader election in IoT networks. In: 15th IEEE Annual Consumer Communications and Networking Conference (CCNC) (2018)
13. Pace, P., Aloi, G., Caliciuri, G., and Fortino, G.: Management and coordination framework for aerial-terrestrial smart drone networks. In: Proceedings of the 1st International Workshop on Experiences with the Design and Implementation of Smart Objects, pp. 37–42, September 2015
14. Murmu, M.K., Singh, A.K.: A bio-inspired leader election protocol for cognitive radio networks. Cluster Comput. **22**(1), 1665–1678 (2018). https://doi.org/10.1007/s10586-017-1677-7
15. Chen, P.Y., Hero, A.O.: Local Fiedler Vector centrality for detection of deep and overlapping communities in networks. In: IEEE International Conference on Acoustic, Speech and Signal Processing (ICASSP) (2014)
16. Arnaud, C.: The Jbotsim Library. HAL Archives Ouvertes (2013)
17. Pushya, C., Aparna, R.A., SivaSankarRao, S.: 3-phase leader election algorithm for distributed systems. In: Proceedings of the Third International Conference on Computing Methodologies and Communication (ICCMC) (2019)

18. Roua, E., Maxime, G., Baudouin, D.: A local leader election protocol applied to decentralized traffic regulation. In: International Conference on Tools with Artificial Intelligence (2017)
19. Mariano, M., Fernando, G.T., Adrian, M.D.: Distributed algorithms on IoT devices: bully leader election. In: International Conference on Computational Science and Computational Intelligence (2017)
20. Newport, C.: Leader election in a smartphone peer-to-peer network. In: IEEE International Parallel and Distributed Processing Symposium (2017)
21. Manisha, R., Purushottam, S., Gargi, M.: Novel leader election algorithm using buffer. In: 2nd International Conference on Telecommunication and Networks (TEL-NET 2017) (2017)
22. Yuichi, S., Fukuhito, O., Taisuke, I., Hirotsugu, K., Toshimitsu, M.: Time-optimal leader election in population protocols. In: IEEE Transactions on Parallel and Distributed Systems (2020)
23. Muhammad, N., Fazli, S., Wazir, Z.K., Basem, A., Nasrullah, A.: Well-organized bully leader election algorithm for distributed system. In: International Conference on radar, Antenna, Microwave, Electronics, and Telecommunications (2018)

Author Index

© The Editor(s) (if applicable) and The Author(s), under exclusive license
to Springer Nature Switzerland AG 2022

H. Ragab Hassen and H. Batatia (Eds.): ACS 2021, LNNS 378, p. 143, 2022.
https://doi.org/10.1007/978-3-030-95918-0